Derk Loorbach • Julia M. Wittmayer •
Hideaki Shiroyama • Junichi Fujino •
Satoru Mizuguchi
Editors

Governance of Urban Sustainability Transitions

European and Asian Experiences

Springer

Editors
Derk Loorbach
Dutch Research Institute for Transitions
Erasmus University Rotterdam
Rotterdam, The Netherlands

Julia M. Wittmayer
Dutch Research Institute for Transitions
Erasmus University Rotterdam
Rotterdam, The Netherlands

Hideaki Shiroyama
The University of Tokyo
Tokyo, Japan

Junichi Fujino
National Institute of Environmental Studies
Tsukuba, Japan

Satoru Mizuguchi
Japanese Forum of Environmental Journalists
Tokyo, Japan

ISSN 2199-5508 ISSN 2199-5516 (electronic)
Theory and Practice of Urban Sustainability Transitions
ISBN 978-4-431-55425-7 ISBN 978-4-431-55426-4 (eBook)
DOI 10.1007/978-4-431-55426-4

Library of Congress Control Number: 2016930105

Springer Tokyo Heidelberg New York Dordrecht London

Printed on acid-free paper

Springer Japan KK is part of Springer Science+Business Media (www.springer.com)

Foreword

The twenty-first century is the Urban Age. Urban areas host more than 50 % of the world's population and produce over 60 % of the economic output as well as of the greenhouse gases emissions globally. These numbers are expected to increase further until the end of the century. It is our challenge to make urban areas more habitable and humane by providing every citizen with the opportunity to work and enjoy their lives while decreasing the environmental burden.

Urban areas are also mega social systems. Mega systems are characterized by huge social inertia. The inertia is constituted of ideals and rules of the past, day-to-day routine, cultures, technologies, and social groups. Transforming urban mega systems into sustainable ones is our challenge of today. It is a formidable task because transformations always include struggling with inertia and resistance from society. The nature of social systems is conservative, inclined to choose the status quo even though in need of change.

Meanwhile, the urgency for change is pressing. IPCC's fifth assessment report explains that a 40–70 % reduction of greenhouse gas emissions (compared with 2010) is necessary to avoid dangerous climate change impacts by the year 2050. The report also illustrates how transforming energy and land use would make it possible to profoundly cut greenhouse gases. Moreover, the report added that we already have examples and methods for transforming finance and governance.

However, transformation is hard to accomplish due to the inertia of social systems. The Japanese government has been developing programs to promote low carbon economy and innovation in urban areas. Two examples are the Eco-Model City from 2008 and the "FutureCity" Initiative from 2012, both of which I have been chairing. Through conducting two programs, we have been experiencing trials and errors in addressing the huge organizational inertia.

In 2012, the second International Conference of the "FutureCity" Initiative invited Derk Loorbach and Julia Wittmayer to lecture on the theory and practice of transition management in Europe. I was convinced that this method can change social inertia through analyzing what it is constituted off, organizing transition arenas by selecting frontrunners with alternative perspectives and practices, shielding the arena from the mainstream, and practicing small-scale social experiments. If this is going well, we can expand it to another arena.

Through promoting the "FutureCity" Initiative, we also have been practicing a somewhat similar method partially without theorizing the process. Derk, Julia, and their colleagues have continued "learning by doing" and "doing by learning" to formulate theories for more than 10 years. In 2014 we decided to work together. This book is the first fruit of our collaboration between Europe and Japan.

The East and the West differ on how to run society. However, we came to know that we also have many similarities. Through our collaborative learning, I expect we can explore new horizons of urban sustainability.

Tokyo, Japan Shuzo Murakami
June 2015

Foreword

Exploring and developing urban sustainability is not only a necessity and moral duty, but also an exciting journey into the future. Traditionally, governments have played a major role in this, and local governments are no exception. Local policies have set ambitious goals, developed subsidies and implementation plans, and facilitated bottom-up initiatives. But as the worldwide population and its consumption, the number of people moving to cities, and the complexity and interconnectedness of our world increases, we need to rethink how we achieve urban sustainability. We need to shift from optimizing the existing solutions to developing completely new ones.

The city of Ghent and its inhabitants are active at the forefront of urban sustainability transitions. We have long acknowledged the need for collaboration between policy, science, and civil society to create a city that is liveable, green, and prosperous. We have gone through historical transitions moving from an industrial city to the current attractive historic and cultural one. Over the last years, we have taken up the challenge to reinvent how we work together to address future challenges. Inspired by the MUSIC project and the transitions approach, a number of our civil servants and social entrepreneurs have started to plant the seeds of transformation.

The way Ghent has chosen is not always the easiest way. Striving for climate neutrality is a challenge in our historic city and vivid harbor. The enthusiasm, leverages, and creativity of all stakeholders are needed to reach this goal in 2050. Ghent strives not only for structural solutions with an impact in the long term, but also chooses a social climate policy. In our policy to make citizens, organizations, institutions, and companies more resilient to rising energy prices by structural energy-saving measures, we have made additional resources available for vulnerable families. Plenty of space has been created in recent years for initiatives from citizens, organizations, and companies to reduce our energy consumption and make our energy production more sustainable. We want to build on this momentum in the coming years because we are convinced that a climate-neutral Ghent is possible only with collaboration. In Ghent, we make this journey with our heads, hearts, and hands. We think, reflect, and rethink where we are and where we want to be. We build on the passion, entrepreneurial spirit, and persistence of change makers across the city. And we choose to do so pragmatically, to start, learn, and improve along the way.

One of the inspiring examples of the impact of our transition process in Ghent is the 'Leefstraat' (in English: *Livingstreet*): for one month a year inhabitants can have their street closed off for cars and reclaim the public space. This experiment where bottom-up meets top-down has since its introduction inspired a growing number of such streets and is now diffusing to other cities as well. Experiments like this can emerge only when the context is right, and I see here a critical role for local policy: bringing together actors to reinvent and remake the city by creating the necessary mental, financial, physical, and regulatory space for sustainability transition. I hope this book gives inspiration and confidence to many more cities to take up the challenge and join this common journey.

Ghent, Belgium Tine Heyse
June 2015

Preface

Both in Europe and in Japan, there are a multitude of reasons for pro-actively seeking strategies for sustainability transitions: ageing populations, economic restructuring, and stagnation under globalization, resources issues, and general liveability. This book aspires to show how accelerating urban sustainability transitions work in practice drawing upon empirical case studies into application of transition management in European and Japanese cities. This volume combines in-depth case studies of transitions in urban contexts, descriptions of successful and failed attempts to manage urban sustainability transitions, and reflections upon the implications and lessons of these. It offers insights, tools, and experiences of a group of frontrunner cities (Aberdeen, Ghent, Higashiomi, Kitakyushu, and Montreuil) in their quest for addressing current unsustainability challenges and move towards a more sustainable future. Building upon exchanges between action researchers, academics, and practitioners from Japan and the Netherlands, we have composed this volume. It shows the diversity of the urban challenges, differences in cultures, and variety of instruments and concepts. But it has also confirmed that there is more that is shared: a language of transformation, a hopeful eye on social innovation and the future of cities, and a dedication to experimentation, learning, and sharing. It is through these mechanisms that we and hopefully those who recognize and feel empowered by our stories will keep transforming cities.

This book is structured in three main parts. Part I, "Introducing Urban Sustainability Transitions and Their Governance", prepares the ground by introducing the city as locus and actor in sustainability transitions (Chap. 1), transition management as a new mode of governing these (Chap. 2) and the networking activities between cities supporting them in becoming more sustainable (Chap. 3). Part II, "Transition Management in Japanese and European Cities", is the heart of the book and includes five rich empirical chapters describing cities and their quest for sustainable development. What holds these chapters together is their engagement with the transition management approach. The chapters on the Japanese cities describe urban governance analysed through a transition management lens, and the chapters on the European cities outline the processes and outcomes of an operational application of transition management, where co-creative processes between city administration and other urban actors were organized based on a transition management methodology. All chapters are structured along a basic rationale, introducing basic background information on the city and its main sustainability challenge, the main analytical framework and/or focus followed by a description of the governance mechanisms addressing the challenge and its outcomes, and closing with an analytical discussion and conclusion. The third and last part of the book, Part III, "Synthesis and Reflections", includes three synthesising chapters. The first of these chapters synthesizes insights and lessons from the experiences with the (operational and heuristic application of the) transition management framework in the five European and Asian cities (Chap. 9). The last two chapters focus on two different audiences: while Chap. 10 draws recommendations for practitioners on the governance of urban sustainability transitions, Chap. 11 draws up a research agenda for this topic.

Part I: Introducing Urban Sustainability Transitions and Their Governance

In the following we outline the main foci of each chapter. After having laid the ground for the series as well as for this book in Chapter 1, the second chapter by Wittmayer and Loorbach is dedicated to an introduction of transition management and its application in the urban context. The authors introduce the principles of transition governance, their translation in a management framework, and the operationalization thereof in instruments and process methodologies. Next to outlining these elements of transition management, they also investigate how these have been applied as a heuristic to analyse urban governance dynamics as well as an operational framework to influence and accelerate sustainability transitions in this urban context. The authors conclude by synthesizing the promises and challenges of transition management in cities and zoom in on the meaning of the urban context for transition management processes.

Chapter 3 describes a number of national and international networks that support cities in working towards a sustainable future for their citizens. In doing so, Wittmayer, Mizuguchi, Rach, and Fujino introduce networks of which the cities featured in this book were part. The focus is on describing the funding context of the networks, their underlying rationale and their role in supporting the cities in their sustainability ambitions. This includes the MUSIC project and network for the European cities and three networks for the Japanese cities, namely, the Eco-Model City, the "FutureCity" Initiative, and the Green and Local Autonomy Model City. This chapter also gives an overview and differentiation of the cities that the book will turn to in the following chapters.

Part II: Transition Management in Japanese and European Cities

In the first empirical chapter (Chap. 4), Frantzeskaki and Tefrati introduce us to the city of Aberdeen, United Kingdom, and its use of a transition management-based co-creative process to address the social and financial vulnerabilities of a new transition away from an oil-dominated economy. In their chapter they focus on the envisioning phase as the "critical phase" in which diverse interests, perspectives, hidden assumptions, and ingrained perspectives are expressed, negotiated, and debated.

The second empirical chapter focuses on Higashiohmi, Japan (Chap. 5). Mizuguchi, Ohta, Beers, Yamaguchi, and Nishimura analyse a community business project called the "Welfare Mall", which clusters local production of food, energy, and elderly care in one location. This chapter focuses on the interactions among these multiple niche-innovations and multi-regimes which occur under the pressure of slow but drastic landscape changes.

Ghent, Belgium, has started up a "climate arena" based on transition management to work towards its ambition of climate neutrality by 2050 (Chap. 6). Hölscher, Roorda, and Nevens focus on the empowerment aspect of a transition management process for policy makers and urban actors involved. It seems to be effective in introducing policy officers to more open and co-creative approaches as well as to cross-departmental collaboration. The challenges regarding the longevity of empowerment effects for other participants are translated into a number of lessons for designing transition management processes.

Shiroyama and Kajiki describe the transition of Kitakyushu, a city infamous for its industrial pollution, towards a focus on environmental conservation, the promotion of environmental business, and welfare for local citizens (Chap. 7). They use transition management as a lens to analyse the Kitakyushu Eco-Town Project, an initiative aimed at promoting environmental business which can be viewed as a catalyst for Kitakyushu's transition to a green city. They pay attention to the important roles of incumbent players, namely, insiders with new ideas in established private and public institutions, for facilitating the transition.

The third European city is Montreuil, France, where a transition management process took place as part of the elaboration of a local climate plan (Chap. 8). Krauz describes the adaptation of transition management to the French political context and explores the possibility of "hybridizations" between "regime" and "niche" actors; between their roles, relations, and cultures.

Part III: Synthesis and Reflections

In the last part of the book, Chap. 9 focuses on synthesizing the main insights and lessons for transition management as a governance approach for urban sustainability transitions. Wittmayer analyses the major challenges cities face in reorienting urban development in a sustainable direction and the potential of transition approaches to contribute to this radical reorientation. The insights focus on the different governance activities and instruments and their usefulness in actually creating space for alternatives and challenging the status quo. The chapter concludes by drawing lessons for the governance principles underlying transition governance as well for accompanying operational process frameworks.

Chapter 10 takes stock of the experiences of the five European and Japanese cities to draft recommendations for practitioners to more effectively deal with urban sustainability issues. Beers reflects on examples of when and why to use transition management specifically. He highlights the importance and potential of network hybridization: the fact that information provided by scientists can act as a common starting point and that striving for shared actions and connecting different problem orientations is more fruitful for transitions than striving for consensus.

Finally, Chap. 11 outlines a research agenda on the governance of urban sustainability transitions building on the discussion and insights from the different chapters. Frantzeskaki and Shiroyama outline three research directions along the different application types of transition management: theoretical, heuristic, and operational applications to advance the research on transition management specifically, and the governance of urban sustainability (transitions) in general.

Rotterdam, The Netherlands — Derk Loorbach
Rotterdam, The Netherlands — Julia M. Wittmayer
Tokyo, Japan — Hideaki Shiroyama
Tokyo, Japan — Satoru Mizuguchi
Tsukuba, Japan — Junichi Fujino

Contents

Part I
Introducing Urban Sustainability Transitions and Their Governance

Part I prepares the ground by introducing the city as locus and actor in sustainability transitions (Chap. 1), transition management as a new mode of governing these (Chap. 2), and the networking activities between the cities featured in this volume supporting them in becoming more sustainable (Chap. 3).

Chapter 1
The Challenge of Sustainable Urban Development and Transforming Cities

Derk Loorbach and Hideaki Shiroyama

Abstract Our quickly changing world faces great challenges when it comes to the sustainable provision of energy, food, shelter, water and welfare to a growing urban population. These grand challenges are increasingly taken up by cities that become the places where sustainable futures are emerging. This chapter introduces the theoretical and practical transition perspective taken in this book and describes its structure and outline. It frames the dynamics in urban development from the perspective of sustainability transitions: deep systemic transformations that are the result of destabilising unsustainable 'regimes' and emerging sustainable 'niches,' driven by transformative agencies and networks. This perspective highlights on the one hand the complexities, uncertainties, and resistance that come along with urban transitions as well as the mechanisms and patterns that enable and accelerate them, and provides the basis for new types of governance. We then describe the structure of the book. It first elaborates upon the theoretical ideas and governance approaches related to sustainability transitions. It then draws upon empirical evidence from applied transition management in European and Japanese cities. In the final part of the book, the authors reflect upon these experiences, to what extent they are comparative, and what can be learnt in general with regard to implementing urban transition strategies.

Keywords Cities • Urban sustainability • Sustainability transitions • Governance

D. Loorbach (✉)
Dutch Research Institute For Transitions (DRIFT), Erasmus University Rotterdam, Rotterdam, The Netherlands
e-mail: loorbach@drift.eur.nl

H. Shiroyama
The University of Tokyo, Tokyo, Japan
e-mail: siroyama@j.u-tokyo.ac.jp

D. Loorbach et al. (eds.), *Governance of Urban Sustainability Transitions*, Theory and Practice of Urban Sustainability Transitions,
DOI 10.1007/978-4-431-55426-4_1

1.1 Urban Sustainability

It is increasingly argued that cities are at the frontline of sustainability: not only are cities most vulnerable to ecological, socioeconomic, and political crises, but they are also hotbeds of innovation and experimentation. Cities are entities in transition themselves as much as that they are the spaces within which novelties emerge. Cities themselves, however, do not constitute a monolithic agency: rather, they are composed of various agents, the organisations and the networks that build and make cities by continuously reshaping, redefining, and reenacting the urban milieu. The greatest part of all the daily actions, innovations and decisions, policy measures, and business strategies, aim to improve existing urban fabrics, economies, and (infra)structures. But the collective impact of urban life is persistently unsustainable, creating negative ecological impacts, social tensions and economic crises. In spite of decades of attention to sustainability, human development is locked into an inherently unsustainable pathway. Taking the perspective of sustainability transitions (Grin et al. 2010), it is argued that deeper and more fundamental shifts are required towards different cultures, structures, and practices that are *inherently sustainable* rather than *less unsustainable*.

There are increasingly strong signals of the emergence of such deeper changes. Sustainable technologies are maturing after decades of experimentation. Across the globe 'translocal' networks of transformation are emerging, developing inherently sustainable alternative systems such as complementary currencies, renewable energy and food collectives and transition towns. Driven by the possibilities of the internet, the availability of open access data and knowledge, and with transformative entrepreneurial capacities, individuals all over the world are linking up in transformative networks. This emerging network society is changing the world collectively in a fragmented, decentralised, and self-organising way—without global negotiations, local agendas, or top-down planning, but through dedication, creativity, persistence, patience, hard work, failure, and recovery. We argue that this transformative human energy could be the most promising development to accelerate towards sustainability. But also that the modern cities with their top-down institutions, hard infrastructures, unsustainable energy systems, dependence on external resources, and unsustainable levels of consumption will not give way easily. This is what we call transforming cities: the uncoordinated yet globally emerging movement in cities where innovative new sustainable solutions are being experimented with, shared, scaled and translated at increasing speed.

Confronting this new reality requires new approaches to governance and change, to help accelerate and guide these emerging sustainability transitions. As we have known for decades, our current pathway of development is not sustainable, and regular policy so far seems unable to shift course. Developed countries are increasingly crossing the ecological, social, and economic boundaries within which production and consumption patterns can be sustained, leading to a variety of problems, crises, and tensions. Rather than addressing these tensions through efficiency increases, technological innovation, and regulatory interventions, transition studies (Rotmans et al. 2001; Grin et al. 2010) suggest that inevitably more unpredictable and chaotic structural changes will take place. As much as such transitions could lead to less desirable futures and collapse, such transitions also offer the possibilities for the relatively rapid breakthroughs that are deemed necessary to achieve global sustainability goals. Actively anticipating and adapting to the dynamics of transitions, transition management (Loorbach 2010) is therefore considered as a way to increase the chances for sustainability transitions.

1.2 Sustainability Transitions in Urban Areas

In this millennium, the global urban population exceeded the rural population for the first time in history (Seto et al. 2010); at present, more than half the world's total population lives in cities (Crosette 2010). As a consequence of the global urbanisation trend, the greatest demand for energy, food, water, buildings, waste management, healthcare, education, and other basic services is concentrated in and around cities. Along with the problem of satisfying increasing demand, through the unwanted by-products of the unsustainable sociotechnological systems in place, cities are responsible for the largest proportion of environmental impacts. Worldwide, cities are responsible for almost 75 % of total resource consumption (Madlener and Sunak 2011) and the primary source of greenhouse gas (GHG) emissions (Grimm et al. 2008). Overall, cities account for at least 70 % of energy-related GHG emissions. Sustainability problems can be found in almost all sociotechnological systems needed to "run cities." As such, cities are the locations where most of the (un)sustainability issues find their origin. In contrast, cities are also locations for sustainability innovation and societal progress; cities can even be considered as potential 'drivers' for sustainable development (Rotmans et al. 2000) or 'hubs' for radical sustainability innovation (Ernston et al. 2010; Bulkely et al. 2012). Instead of seeing cities as centralised bureaucracies or static entities, we follow the more recent conceptualisation of cities as multifaceted 'municipalities' behaving as self-governing entities on sustainability issues (Burstrom and Korhonen 2001). In this view, cities take the lead in sustainability solutions. And although they might not be the exclusive locations to advance sustainability transitions, cities can at least play an important role on two levels: as 'actors' with regard to (re)developing sociotechnological systems and as facilitators of locations for sustainability innovations (Geels et al. 2011). Agents, in general, can "push policy" towards promoting sustainability on the street level, along with (re)structuring the city's infrastructure, and facilitating larger-scale environmental, social, and economic innovations throughout the world.

This insight that cities are 'actors and locations' of sustainability transition is not new. Many ambitious sustainability initiatives have already emerged at the level of cities and metropolitan regions, such as the Covenant of Mayors and C40 climate coalition. From a transitions perspective, we argue that although indeed the potential for accelerating towards sustainability is there, cities are not automatically able to proactively anticipate and adapt to such possibilities. More important, when cities do take up their role as actors, the available 'development' policies so far seem insufficient to guide and accelerate deeper systemic change towards sustainability. Although development policies have added to economic and technological improvements in many cities around the world, simultaneously global consumption and the waste and emission levels stemming from these improvements have continued to increase. In this view, regular urban policies and cities' governance structures focus too much on "straightforward" economic development and standard technological solutions, whereby adaptive and transformative strategies for sustainability are needed. Sustainability is too often considered as a separate domain of secondary policy concern, mainly because it is dominantly perceived in the realm of short-term economic calculus. Thus, its advance is perceived as very costly and uncertain. Clearly, it takes massive investments to (re)structure cities' sociotechnological systems towards sustainable functioning. Still, the transition perspective suggests that no matter the high costs and level of uncertainty, the costs of inaction are in the longer run always higher. Because sociotechnological systems are embedded in societies, which in turn are embedded in their environment (Giddings et al. 2002), the persistent unsustainability inevitably will lead to a deep crisis in the current systems and their possible collapse. The primacy of short-term economic concerns in policy making is shortsighted, as the operation of economic systems is dependent on the sustainable functioning of sociotechnological systems as well as their societal and environmental surroundings.

Awareness of the unsustainability of our current developments pathways reached the global stage with the introduction of the notion of *sustainable development* by the so-called Brundtland commission in the 1980s; they defined it as "Meeting the needs of the present without compromising the ability of future generations to meet their own needs" (WCED 1987). The report concluded, in line with studies such as the "Limits to Growth" report, that there is an inherent unsustainability in the dominant development pattern. We are still locked into an unsustainability pathway, even after three decades of research, policy, and debate seeking to define, operationalise, and implement sustainable development. It can therefore be argued that sustainable development in its current institutionalised forms has become part of sustaining systemic unsustainability (Loorbach 2014) and that we need to fundamentally reconsider this notion and how it should be operationalised in practice. To do so, we start from identifying common ground as well as seek to identify seeds of transformative social change towards inherent sustainability rather than incremental change leading to reduced unsustainability.

A common ground is found in that it encompasses social, environmental, and economic prosperity, here and in other places, now and in the future. What this exactly means differs from situation to situation, as the needs of people vary, depending among others on historical, political, economic, social, and ecological circumstances and developments. The many forms of unsustainability in Western societies are visible in the form of what we call persistent problems. Examples of such problems at different scale levels are climate change at a global level; the agricultural problem at a continental scale, with animal diseases such as bird flu, mad cow disease, and foot-and-mouth disease; and the mobility problem at a national level with traffic congestion and air pollution from increased mobility (Rotmans et al. 2007). These problems are complex, as they are deeply embedded in societal structures and institutions. They have multiple causes and consequences, and their reach extends beyond a wide range of societal domains, people, and scale levels.

A clear example of this is the unsustainability of our current energy systems and how this impacts urban life. Most modern cities that historically had their own energy supply now depend on national energy grids and power plants running on fossil fuels in a liberalised market. Cities experience the effects of pollution, price volatility, and dependence upon foreign supply but often do not have the position or instruments to change this. Alternatives have for long been considered as politically contested or too small in scale to replace current options. The interests, investments, and stakes in the current fossil-based regime are high, so that structural change cannot be expected to be initiated by the vested parties, also called regime, which is defined as the dominant constellation of structures, culture, and practices in a certain societal system. But now many cities are putting substantial efforts into developing new urban energy solutions, which range from stimulating energy efficiency, to diffusing renewable technologies, to developing waste-heat systems or smart grids. Cities are in this way providing space for radical alternatives to the dominant fossil fuel-based and centralised energy system, thereby becoming important change agents in transitions. It is, however, in practice a rather chaotic and uncoordinated process in which different cities explore different strategies, solutions, and technologies, depending on their local context, challenges, and potential. In terms of transition governance, developing a sustainable energy system thus is not a process of planning and control but requires clever organisation and facilitation through creating room for self-organisation, experimentation, and learning.

A transition, then, is a structural change in a societal system or subsystem that is the result of a co-evolution of economic, cultural, technological, ecological, and institutional developments at different scale levels. A transition consists of a number of system changes, which are organisation-transcending innovations that fundamentally alter the relationships between companies, organisations, institutions, and individuals in a certain field or domain (a subsystem). Transitions at this societal level could take up to two generations to materialise and thus require concerted efforts that go beyond the time horizon and possibilities of individual organisations or even the government alone. To direct transitions towards sustainability, new modes of governance are needed that take into account the long time horizon, the uncertainties and complexities, and the multitude of persons and interests involved. Transitions therefore imply different roles and practices from individual factors involved, such as companies, scientific institutes, governmental organisations, or NGOs.

Transition studies as a scientific field have during the past years developed five important perspectives for analysing sustainability transitions in urban areas:

1. *The niche perspective:* a focus on microlevel innovations that have the potential to radically change the urban fabric and social practices towards sustainability even when these changes are costly, novel, and spatially segmented.
2. *The multi-phase perspective:* a holistic and dynamic understanding of the multiple phases (i.e., pre-development, take-off, and lock-in) and the associated dynamics that a transition process can display.
3. *The co-evolution perspective:* the conceptual tools to understand what contributes to evolutionary interactions between environment and societal transformations happening over a long period of time in an incremental way.
4. *The multi-pattern perspective:* the different patterns of processes in which transitions can proceed when considering policy, institutions, technology, and agency dynamics.
5. *The multi-level perspective:* the different levels of dynamics in which interlevel and intralevel interactions influence the transition as a whole.

In combination, these five (non-exclusive) perspectives make up the basis for transition analyses and provide valuable insights when trying to discover possible management strategies for transitions in specific urban areas. One of the key characteristics of the transition perspective is that it focuses on the niche experiments and incremental multisectoral co-evolution rather than transformation based on top-down initiative. The perspectives are also based on generic concepts, which allow for adaptation in context-specific circumstances. The concepts of niches, multi-phases, co-evolution, multi-patterns, and multi-levels acquire different meanings along with the characteristics of the sociotechnological and societal system under study (Frantzeskaki and Loorbach 2010), which makes them applicable in various situations (e.g., resource scarcity, pollution, climate change, loss of biodiversity, and waste management), especially those wherein the transition dynamics are still developing and underdetermined at the time. The five perspectives provide concepts for innovation and "stepping stones" for crossing theoretical gaps to an adaptable framework for proactive transition management of specific sustainability problems in a certain urban area.

1.3 Governance of Urban Sustainability Transitions

Cities are perhaps the most suitable level of application of transition management in many ways: because of the multiple dynamics occurring at this level and the concrete level of application, but also because of the close relationship between those participating in transition programs, transition arenas, and transition experiments. It is argued that cities are simultaneously the places confronted with the impacts of unsustainability and therefore also a proper context for experimentation with new solutions. In that sense cities are 'transition machines,' producing the innovations driving and accelerating larger-scale transitions. It is a level where results can be concrete and accessible to a broader group: if these are missing, this will also not go unnoticed. On the other hand, we need to remain critical whether cities are the actual level where a lever for system change can be enacted.

The transition management perspective suggests that, in cities, a number of (often interwoven) economic, technological, and institutional barriers exist to act and invest in sustainability innovation. These barriers create a lock-in not only of unsustainable systems already in place but also in the problem solving of future developments. One of the main characteristics of this lock-in is the managerial focus on merely addressing sector-specific and "manageable" problems through formalised policy-making processes (Loorbach and Rotmans 2010). From a transition management perspective, we perceive cities as complex adaptive systems; namely, that to a large extent cities are self-organising systems with emergent properties and adaptive capabilities that add to the sustainability problems or both solutions. Therefore, they can defy tightly structured top-down control (Nevens and Roorda 2014). In this view, the problems that regular policies try to solve, such as pollution, emissions, and congestion, are often only the symptoms of underlying systemic problems.

The complexity and persistence of the challenges we are facing are indeed pervasive and severe at a city level. But here we should not disregard the city's global impact. To contribute to solving sustainability issues, the view on cities needs to transcend the (perceptive) barriers of spatial scale, because global issues do not immediately relate to action at the street level. Global sustainability problems caused on the city level imply that (multi)national, regional, and local governance structures cannot tackle them alone. The different levels of scale require a multiscale approach that can zoom in and out from the macro- to the meso- to the microlevel, and vice versa. Solutions require broadly carried "bottom-up" initiatives and innovations that can connect and interact with governance structures and "top-down" policies on higher levels. In this way, cities can act as a facilitator of sustainability transitions, in terms of committing to long-term investments in sustainable sociotechnological (infra-)structures that create space for the emergence of more sustainable alternatives (Loorbach et al. 2010). However, because global issues go beyond classic terms of local policy cycles, and the long-term investments span multiple terms (perhaps even generations), the temporal scale (i.e., "not in my term") of policy making should be extended. In addition, the institutional scale should be more prominently addressed, as global problems are easily considered 'not my personal business,' and local autonomy is often too restricted by central governance to act effectively against global problems on a local scale (Bai 2007).

This is where the approach of transition management, translated to the urban context, is increasingly seen as promising to help guide and accelerate the emerging transitional dynamics in cities (see Wittmayer and Loorbach, Chap. 2, this volume; Frantzeskaki and Loorbach 2010; Frantzeskaki et al. 2012; Jefferies and Duffy 2011; Loorbach et al. 2009; Vergragt and Brown 2010; Wittmayer et al. 2015; Nevens and Roorda 2014). The transition management approach provides a number of basic governance starting points, a governance framework and specific policy instruments applied in the specific context of urban areas. It provides a way to analyse, describe, and reflect upon emerging transformative governance processes and networks, and as well it can be used to proactively develop transition arenas, experiments, and network building on present activity. Although in its early years transition management (Rotmans et al. 2001; Loorbach 2007) was mainly applied on a sector level or a (subnational or) regional scale, cities seem perhaps an even more natural context in which to apply transitionmanagement.

Since 2001, experiments have been emerging in the Netherlands (and to some extent in other Western European countries) with the approach of transition management. Basically, this approach starts from the conceptualisation of structural societal change as a transition: a long-term, multilevel process of change in which distinctive phases of changes can be distinguished. The transition concept (Rotmans et al. 2001; Geels 2002; Berkhout et al. 2004; Elzen et al. 2004; Meadowcroft 2005; Van der Brugge and Rotmans 2005), which has been evolving over time, is used to analyse and understand the dynamics of structural change in societal systems. Based on insights in the dynamics of change, transition management offers a basic starting point for influencing the speed and direction of such ongoing transitions towards sustainability.

Key elements in the transition management approach are frontrunner networks that develop an integrated understanding of their common transition challenge and a desirable future perspective; a shared transition agenda as a roadmap for social innovation; transition experiments as innovation icons to implement parts of the future agenda; and monitoring, evaluation, and adaptation. By together developing such activities and constantly deepening the collective understanding of the societal transition of which the actors are part, reflexive and strategic capacity is being built up within an evolving network. Over time, this enables participants to engage in direct competition with existing regime actors and networks.

The point of transition management is that urban governance—from the (inter) national to the street level—can only advance a global sustainability transition by developing alternative means, strategies, and instruments that acknowledge the problems that come along, managing highly complex processes on different scales, but also the opportunities present in the self-organising dynamics. Transition management experiments show that such "new modes of governance" can influence the speed and direction of transitions in a subtle and indirect way, namely by facilitating location and stimulating actor dynamics across sectors that move towards sustainability, and in the process come up with innovative ways to get there eventually. These dynamics include different perspectives, addressing several aspects of the transition in multiple phases, domains, and on multiple levels. It is also important to note that tensions in incumbents can be drivers for transition, synchronising with the dynamics in wider society.

References

Bai X (2007) Industrial ecology and the global impacts of cities. J Ind Ecol 11(2):1–6

Berkhout F, Smith A, Stirling A (2004) Socio-technical regimes and transition contexts. In: Elzen B, Geels FW, Green K (eds) System innovation and the transition to sustainability. Edward Elgar, Cheltenham

Bulkely H, Broto VC, Edwards G (2012) Bringing climate change to the city: towards low carbon urbanism? Local Environ 17(5):545–551

Burstrom F, Korhonen J (2001) Municipalities and industrial ecology: reconsidering municipal environmental management. Sustain Dev 9(1):36–46

Crosette B (2010) State of world population 2010: from conflict and crisis to renewal. United Nations Population Fund, New York

Elzen B, Geels FW, Green K (eds) (2004) System innovation and the transition to sustainability. Edward Elgar, Cheltenham

Ernston H, Van der Leeuw S, Redman C, Meffert D, Davis G, Alfsen C, Elmqvist T (2010) Urban transitions: on urban resilience and human-dominated ecosystems. AMBIO 39(8):531–545

Frantzeskaki N, Loorbach D (2010) Towards governing infrasystem transitions: reinforcing lock-in or facilitating change? Technol Forecast Soc Chang 77(8):1292–1301

Frantzeskaki N, Loorbach D, Meadowcroft J (2012) Governing transitions to sustainability: transition management as a governance approach towards pursuing sustainability. Int J Sustain Dev 15(1/2):19–36

Geels FW (2002) Understanding the dynamics of technological transitions: a co-evolutionary and socio-technical analysis. Universiteit Twente, Enschede

Geels F, Kemp R, Dudley G, Lyons G (2011) Automobility in transition? A socio-technical analysis of sustainable transport. Routledge, New York

Giddings B, Hopwood B, O'Brien G (2002) Environment, economy and society: fitting them together into sustainable development. Sustain Dev 10(4):187–196

Grimm NB, Faeth SH, Golubiewski NE, Remand CL, Wu J, Bai X, Briggs JM (2008) Global change and the ecology of cities. Science 319(5864):756–760

Grin J, Rotmans J, Schot J, Geels F, Loorbach D (2010) Transitions to sustainable development. Part 1. New directions in the study of long term transformative change. Routledge, New York

Jefferies C, Duffy A (2011) The SWITCH transition manual. University of Abertay, Dundee

Loorbach D (2007) Transition management. New mode of governance for sustainable development. PhD thesis. Erasmus University, Rotterdam

Loorbach D (2010) Transition management for sustainable development: a prescriptive, complexity-based governance framework. Governance 23(1):161–183

Loorbach D (2014) To transition! governance panarchy in the new transformation. DRIFT/EUR, Rotterdam

Loorbach D, Rotmans J (2010) The practice of transition management: examples and lessons from four distinct cases. Futures 42(3):237–246

Loorbach D, Frantzeskaki N, Meadowcroft J (2009) Discovering sustainability: a transition approach towards sustainable development. IHDP conference on governance and sustainability. Bonn, Germany

Loorbach D, Frantzeskaki N, Thissen W (2010) Introduction to the special section: infrastructures and transitions. Technol Forecast Soc Chang 77(8):1195–1202

Madlener R, Sunak Y (2011) Impacts of urbanization on urban structures and energy demand: what can we learn for urban energy planning and urbanization management? Sustain Cities Soc 1(1):45–53

Meadowcroft J (2005) Environmental political economy, technological transitions and the state. New Polit Econ 10(4):479–498

Nevens F, Roorda C (2014) A climate of change: a transition approach for climate neutrality in the city of Ghent (Belgium). Sustain Cities Soc 10:112–121

Rotmans J, Kemp R, Van Asselt M, Geels F, Verbong G, Molendijk K (2000) Transities & transitiemanagement: de casus van een emissiearme energievoorziening. ICIS/MERIT, Maastricht

Rotmans J, Kemp R, Van Asselt M (2001) More evolution than revolution: transition management in public policy. Foresight 3(1):15–31

Rotmans J, Loorbach D, Kemp R (2007) Transition management: origin, evolution, critique. DRIFT/Erasmus University Rotterdam, hdl.handle.net/1765/37240

Seto KC, Sánchez-Rodríguez R, Fragkias M (2010) The new geography of contemporary urbanization and the environment. Annu Rev Environ Resour 35:167–194

Van der Brugge R, Rotmans J (2005) Towards transition management of European water resources. Water Resour Manag 21(1):249–267

Vergragt PJ, Brown HS (2010) Managing urban transitions: visioning and stakeholder collaboration. A case study in transforming residential housing in Worcester. George Perkins Marsh Institute, proceedings of Sussex Energy Group conference: energy transitions, Sussex

WCED (1987) Our common future. University Press, Oxford

Wittmayer JM, Rok A, Roorda C, van Steenbergen F (2015) Governing sustainability: a dialogue between Local Agenda 21 and transition management. Local Environ. Online first. doi:10.1080/13549839.2015.1050658

Chapter 2
Governing Transitions in Cities: Fostering Alternative Ideas, Practices, and Social Relations Through Transition Management

Julia M. Wittmayer and Derk Loorbach

Abstract Sustainability transitions pose novel challenges to cities that go beyond traditional planning and urban development policies. Such transitions require broader engagement, empowerment, and breakthrough strategies which enable, facilitate, and direct social innovation processes towards adaptive and innovative urban futures. The transition approach offers a set of principles, a framework, instruments, and process methodologies to analyse as well as systematically organise and facilitate such social learning and innovation processes. During the past decade, researchers and policy entrepreneurs around the world have been experimentally applying the transition perspective in practice under the label of 'transition management'. This approach is based on bringing together frontrunners from policy, science, business, and society to develop a shared understanding of the joint complex transition challenge, to develop collective transition visions and strategies, and to start strategic experiments. In this chapter we zoom in on the different elements of transition management (i.e., principles, framework, instruments, process methodologies) and their heuristic and operational use in the urban context.

Keywords Heuristic • Process methodologies • Sustainability transitions • Transition management • Urban context

2.1 Introduction

When talking about cities and the local level, there is no circumventing the impact that was caused by the 1992 United Nations Conference on Environment and Development in Rio de Janeiro. Here, the local level prominently entered the stage as an important context in which to address sustainability concerns as

J.M. Wittmayer (✉) • D. Loorbach
Dutch Research Institute for Transitions, Erasmus University Rotterdam, Rotterdam, The Netherlands
e-mail: wittmayer@drift.eur.nl

D. Loorbach et al. (eds.), *Governance of Urban Sustainability Transitions*, Theory and Practice of Urban Sustainability Transitions,
DOI 10.1007/978-4-431-55426-4_2

"*so many of the problems and solutions being addressed by Agenda 21 have their roots in local activities*" (UNCED 1992, Agenda 21, Chap. 28). In the decade after, this led to the emergence of thousands of Local Agenda 21 processes addressing sustainability concerns in cities, towns, and neighbourhoods all over the world (ICLEI 2012). Presently, some of these processes still flourish, whereas in Europe most have triggered follow-ups or have died out. The decreasing importance of this specific local process, as well as a more receptive local government sphere, are the backdrops for current ideas and practices of transition governance (Wittmayer et al. 2015).

A number of governance approaches have been developed in the context of a complex and uncertain world facing persistent problems deeply embedded in societal structures and multi-actor contexts. Such approaches aim to address the tension between "*the open-ended and uncertain process of sustainability transitions and the ambition for governing such a process*" (Frantzeskaki et al. 2012b). Examples are ideas and notions about adaptive governance (Olsson et al 2006), reflexive governance (Voß et al. 2006; Grin et al. 2010), or transition governance (Loorbach 2007; Frantzeskaki et al. 2012b). These governance notions address a reality perceived as multiscalar, complex, nonlinear, uncertain, normative, dynamic, complex, and path dependent. From different (multi-)disciplinary backgrounds, these notions have been further developed into more specific approaches, such as empowering designs (Leach et al. 2010), strategic niche management (Kemp et al. 1998; Schot and Geels 2008), and transition management (Rotmans et al. 2001; Loorbach 2010; Frantzeskaki et al. 2012b). This chapter zooms in on transition management as a form of transition governance and specifically focusses on its recent 'urban turn.'

When we refer to the urban context, we focus in particular on a number of specific characteristics of cities that should be taken into account in transition governance—namely, personal, institutional, and geographic proximity—as well as multiscalar and multi-domain interaction (see Table 2.1; cf. Loorbach and Shiroyama 2016, Chap. 1, this volume).

The notion of transition management was developed in the science policy debate leading up to the fourth National Environmental Policy Plan (NMP4) in the Netherlands in 2001 (Rotmans et al. 2001; Kemp and Rotmans 2009; Loorbach and Rotmans 2012; Voß 2014). During the past decade, researchers and policy entrepreneurs around the world have been experimentally applying the transition perspective in practice under the label of 'transition management.' This approach is based on (1) bringing together frontrunners from policy, science, business, and society to develop shared understandings of complex transition challenges; (2) developing collective transition visions and strategies; and (3) experimentally implementing strategic social innovations.

Transition management provides researchers with analytical lenses (i.e., heuristics; see Mizuguchi et al. 2016, Chap. 5, this volume; Shiroyama and Kajiki 2016, Chap. 7, this volume; Frantzeskaki et al. 2014a; Brown et al. 2013) to understand and analyse the dynamics of urban sustainability transitions both historically and in transitions in the making. Its concepts, introduced in more detail next, are also seen as powerful operational tools to help conceptualise and address the fundamental

Table 2.1 Characteristics of the urban context

Characteristic	Description
Geographic proximity	Cities as places where spatial distances are smaller as compared to regions or countries (Boschma 2005; Coenen et al. 2012; Raven et al. 2012)
Multiscalar interaction	Cities as being nested in and constituting of different spatial scales and networks. Scales as actively constructed and interacted with, in ways that support actors in achieving their goals (Coenen et al. 2012; Nevens et al. 2013; Coenen and Truffer 2012)
Multidomain interaction	Cities as places where changes in different domains (e.g., energy, mobility, social care) come together and interact. (Nevens et al. 2013)
Personal proximity	Cities as living environments in which people have personal, emotional, and social stakes, including socially embedded relations and a level of trust (*Related to the concept of social proximity by* Boschma 2005)
Institutional proximity	Cities share formal and informal institutions, including laws and rules as well as cultural norms and habits. (Boschma 2005)

changes necessary to move towards sustainable cities. They help people working on urban development to understand the complexity of their task and the complexity of the system they aim to influence and change. They also support articulating (shared) long-term ambitions to guide short-term actions (see Hölscher et al. 2016, Chap. 6, this volume; Frantzeskaki and Tefrati 2016, Chap. 4, this volume; Krauz 2016, Chap. 8, this volume; Wittmayer et al. 2014a, b; Roorda et al. 2014).

Transition management has been challenged and further developed through theoretical work and heuristic and operational application. Theoretical contributions focus on developing the concept by either grounding it in specific theories (e.g., Rotmans and Loorbach 2009; Frantzeskaki et al. 2012b) or by critiquing specific aspects, most prominently issues of power, politics, and agency. In terms of the latter, much theoretical work as well as practical experimentation sought to deepen our understanding of power relations and political implications and how they could be addressed (Smith et al. 2005; Shove and Walker 2007; Hendriks 2009; Avelino 2009; Kern and Howlett 2009; Meadowcroft 2009; Smith and Stirling 2010; Kern 2012; Jhagroe and Loorbach 2014). These contributions identify challenges of transition management in terms of who is governing, whose framings count (in terms of system, problems, goals, sustainability), and what is the relationship with democratic institutions, incumbent regime actors, and dominant discourses. Many of these challenges and others, such as the narrow focus on desired (versus undesired) transitions, technical systems, and a specific group of key actors, have been addressed in more recent work on transition management (see, for example, the chapters in this volume). Heuristically and operationally, transition management has been applied in a number of functional domains such as energy (Verbong and Loorbach 2012), water (Van der Brugge et al. 2005), and mobility (Avelino et al. 2012). Only quite recently has it been used to describe and prescribe governance processes in geographically bounded systems, such as cities (Nevens et al. 2013; Nevens and Roorda 2014; Ferguson et al. 2013; Wittmayer et al. 2014b, 2015), towns, and urban neighbourhoods (Wittmayer et al. 2014a, b).

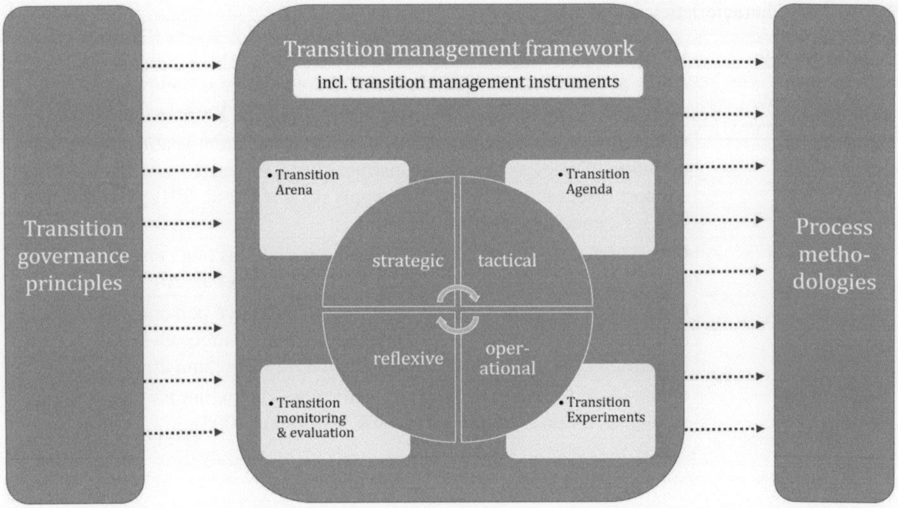

Fig. 2.1 Elements of transition management

After outlining the methodology (Sect. 2.2), we scrutinize transition management in the urban context by outlining different elements thereof and the ways these have been used heuristically and operationally (Sect. 2.3). With elements, we refer to (a) the principles of transition governance, (b) their translation in a management framework, and its associated operationalisation in terms of (c) instruments and (d) process methodologies (Fig. 2.1). Based on this analysis, we synthesise the promises and challenges for making space for alternative ideas, practices, and social relations in cities; and scrutinize the characteristics of the urban context and their meaning for transition management processes (Sect. 2.4).

2.2 Methodology

This chapter is based on both our experience in working with transition management and a literature review of transition management in the urban context. Both authors are involved in the practical and theoretical development of transition management thinking, from the very start of the concept (second author) up to its recent 'urban turn.' Our literature review encompassed more general literature on the theoretical and practical foundations of transition management next to literature on its applications in the urban context. Articles relating to the former were selected based on our experience with the field. These articles are used to provide an overview of the development of transition management, its different elements (principles, framework, instruments, process methodologies), as well as the different critiques it spurred. The literature on transition management in the urban context is just starting to emerge. We could identify a number of relevant articles examining the development, premises, and/or results of transition management in the urban context by using Scopus and snowballing. This sample was broadened by reviewing grey literature on transition management in the urban context such as project reports. For the latter, we mainly focussed on the outputs of two European projects that constituted a breeding ground for the conceptualisation of transition management in the urban context: the FP7-funded InContext project (2010–2013) and the EU-Interreg-funded MUSIC project (2010–2015). As our focus in this chapter is on applications of transition management, we did not include similar developments in transdisciplinary sciences in this review (Wiek 2007; Lang et al. 2012; Wiek et al. 2014).

2.3 Transition Management

In this section we outline the elements of transition management, namely, the principles of transition governance, their translation in a management framework, and its associated operationalisation in terms of instruments and process methodologies (see Fig. 2.1). For each element, we first give a basic description and then show how it has been used in the context of cities, towns, and neighbourhoods. In so doing, we distinguish between different application types of transition management, namely, *heuristicapplications*, employing the elements as an analytical lens for understanding and explaining governance processes, and *operationalapplications*, describing the application of transition management process tools to set up participatory sustainability processes (cf. Frantzeskaki et al. 2014b).

2.3.1 Transition Governance Principles

Since its inception, the concept of transition management as a governance approach to sustainability transitions has been theoretically further developed and grounded in complex systems, governance, and sociological theories (Loorbach 2007, 2010; Rotmans and Loorbach 2009; Grin et al. 2010; Frantzeskaki et al. 2012b). Based on an understanding of transitions as processes of fundamental long-term multilevel and multiphase change in complex, adaptive systems, a number of governance principles have been formulated. Building on work by Kemp and Rotmans (2009), Loorbach (2010, pp. 167–168) outlines the following nine principles for transition management.

- The dynamics of the system create feasible and nonfeasible means for steering: this implies that *content and process are inseparable*. Process management on its own is not sufficient—insight into how the system works is an essential precondition for effective management.
- *Long-term thinking (at least 25 years) is a framework for shaping short-term policy* in the context of persistent societal problems. This concept requires

backcasting and forecasting: setting of short-term goals, based on long-term goals, and reflection on future developments through the use of scenarios.
- *Objectives should be flexible and adjustable at the system level.* The complexity of the system is at odds with the formulation of specific objectives and blueprint plans. While being directed, the structure and order of the system are also changing, and so the objectives set should change too.
- The *timing of the intervention is crucial.* Immediate and effective intervention is possible in both desirable and undesirable crisis situations.
- Managing a complex, adaptive system means *using disequilibria as well as equilibria.* Relatively short periods of nonequilibrium therefore offer opportunities to direct the system in a desirable direction (towards a new attractor).
- *Creating space for agents to build up alternative regimes* is crucial for innovation. Agents at a certain distance from the regime can effectively create a new regime in a protected environment to permit investment of sufficient time, energy, and resources.
- *Steering from 'outside' a societal system is not effective*: Structures, actors, and practices adapt and anticipate in such a manner that these should also be directed from 'inside.'
- A *focus on (social) learning* about different actor perspectives and a variety of options (which requires a wide playing field) is a necessary precondition for change.
- *Participation from and interaction between stakeholders* is a necessary basis for developing support for policies but also to engage actors in reframing problems and solutions through social learning.

Following these principles, transition management clearly perceives the governance of sustainability transitions as an open-ended process of searching, learning, and experimenting within societies. It has a clear focus on innovation and sustainability, because "*todevelop sustainably means tocontinuouslyinnovate and redefine existing culture, structures and practices in an evolutionary manner*" (Frantzeskaki et al. 2012b, p. 25). These principles offer a basic starting point for experimental operationalisation as well as for analysis and reflection.

Initially, these principles have been formulated, as well as further developed and empirically grounded, in the context of functional systems as well as a regional systems (cf. Loorbach 2007) and as such are not specific to the urban context. To date, there has been no reflection or adaptation of these principles to the urban context (Frantzeskaki et al. 2014b). The synthesis chapter of this book, which distils additional principles for transition governance in cities based on insights from this volume, is an exception in this regard (Wittmayer 2016, Chap. 9, this volume).

2.3.2 Transition Management Framework

The rather abstract governance principles have been translated in a management framework, the transition management cycle (see middle part of Fig. 2.1 for a simplified version). This framework distinguishes between governance activities at the following four levels (see Loorbach 2007, 2010).

- *Strategic-level activities*: Activities aimed at the long term through which the future is collectively debated and imagined; for example, visioning, long-term goal formulation, including collective goal setting and norm setting.
- *Tactical-level activities*: Activities aimed at the midterm and long term, targeting changes in established structures, institutions, regulations, and physical or financial infrastructures.
- *Operational-level activities*: Activities aimed at the short term, focussing on experiments and actions through which alternative ideas, practices, and social relations are practised, tried out, and showcased.
- *Reflexive-level activities*: Activities aimed at learning about the present state and dynamics in the system, and about possible future states as well as about the way from present to future: these include (collective) learning from ongoing operational, tactical, and strategic activities.

Although these activities are recognisable in other governance approaches or policy process models, their difference here lies in their focus on societal processes, persistent problems, fundamental change, and innovation as well as their normative direction (i.e., sustainability) (Frantzeskaki et al. 2012b; Loorbach 2010).

This framework has been used as a heuristic in cities to understand and interpret ongoing governance processes. By way of example, Frantzeskaki et al. (2014a) have been using the different governance levels as part of a mapping framework, which they developed to examine the governance imprint of urban partnerships in the redevelopment of the former Rotterdam City Port area along two axes: their impact in terms of synergies and the governance role they adopt. The framework makes it possible to identify agency patterns at different levels: the way these influence and interact with their broader context (i.e., the status quo) and add up to generate movement into a certain direction. From this perspective, each type of governance activity has distinguishable forms of agency, instruments, processes, and organisational logics. The authors conclude that actively seeking to engage with existing forms of transition governance through systematic intervention strategies supports influencing and accelerating transitions. Two contributions of this volume also use the levels of governance activity to reflect on (1) the value of an operational transition management envisioning process (Frantzeskaki and Tefrati 2016, Chap. 4, this volume) and (2) the transition governance activities in Higashiomi and especially the importance of the reflexive activities in realising a multi-niche innovation (Mizuguchi et al. 2016, Chap. 5, this volume).

2.3.3 Transition Management Instruments

This transition management framework (i.e., the transition management cycle) also connects a number of instruments to each of the governance levels. The cyclical nature of the framework implies that strategic-level activities are followed by tactical and operational instruments and closing the cycle with reflexive ones. However, the cycle has to be understood as iterative (Loorbach 2010); activities can be started at each of the governance levels, thus on the operational level rather than on the strategic level, for example (Van den Bosch 2010), and can run in parallel (Wittmayer et al. 2014a). Thus, the activities and instruments interact more than is implied by the following presentation.

On a strategic governance level, the so-called transition arenahas been developed as a process instrument to develop a new narrative and discourse to frame and guide sustainability transitions; this is simultaneously referred to as a setting as well as a "*small network of frontrunners with different backgrounds*" (Loorbach 2010, p. 173). Frontrunners are selected based on their diverse societal values and perspectives and on the alternatives that they offer in terms of ideas, practices, or social relationships with regard to the status quo (Wittmayer et al. 2011). The perspectives of the frontrunners are subsequently confronted and possibly integrated in a participatory learning process (van Buuren and Loorbach 2009). A substantive outcome of the process is a transition narrative for the city, which consists of (a) a shared integral problem statement outlining the need for a transition, (b) a novel future perspective including sustainability criteria, and (c) transition images and pathways. This narrative plays into existing dynamics and discourses and creates alternative futures and discourses aimed at influencing the direction of change. The underlying idea is that this narrative inspires and motivates social innovation and creates a broader movement (Loorbach 2007). In addition, the process of producing the narrative should lead to social and second-order learning, through which participants (i.e., frontrunners) are encouraged to engage in tactical and operational activities, as outlined next.

Tactical governance activities include, for example, dividing the transition narrative in achievable steps or a roadmap, the *transition agenda*. Activities include the exploration of structural barriers through transition scenarios (Sondeijker 2009) or backcasting (Quist et al. 2011, 2013). Backcasting leads to the exploration and framing of specific transition pathways, which are further developed through negotiation, collaboration, and coalition building (Frantzeskaki et al. 2012b). *Transition experiments*, which are considered instruments at operational governance level, are aimed at learning about putting the narrative into practice, possibly along a certain transition pathway. This placement can take place either through conceiving of new alternatives realised through a project structure, or through broadening, deepening, and scaling up existing and planned initiatives and actions (Van den Bosch and Rotmans 2008). As opposed to a regular project, a transition experiment is an "*innovation project with a societal challenge as a starting point for learning aimed at contributing to a transition*" (Van den Bosch 2010, p. 58). Reflexive governance activities take place throughout to evaluate and monitor the transition process and the various levels and their interrelationships as well as the transition management framework itself: this is the reflection part where changes in the urban fabric and dynamic become registered, existing tools are adapted, and new insights are formulated. *Transition monitoring* not only aims at gathering data but also includes intervention on the basis of these data (Taanman 2014).

These instruments have been translated for the urban context in the concept of 'Urban Transition Labs' (Nevens et al. 2013). Inspired by the transdisciplinary living labs approach, the authors "*consider an Urban Transition Lab as the locus within a city where (global) persistent problems are translated to the specific characteristics of the city and where multiple transitions interact across domains, shift scales of operation and impact multiple domains simultaneously (*e.g. *energy, mobility, built environment, food, ecosystems).Itis a hybrid, flexible and transdisciplinary platform that provides space and time for learning, reflection and development of alternative solutions that are not self-evidentin a regime context*" (Nevens et al. 2013, p. 115). This approach promises the creation of a systems thinking mindset, a strategic agenda and related short-term actions, space, and empowerment starting from selective participation, as well as a setting of learning (Nevens and Roorda 2014).

The instruments and the underlying principles of transition management have inspired different developments. By way of example, the City of The Hague, The Netherlands experimented with a new kind of subsidy scheme for creating a climate movement in the city (Avelino et al. 2011; Wittmayer 2014). Also, the Japanese *"Future City" Initiative* has been inspired by the transition management approach (see Wittmayer et al. 2016, Chap. 3, this volume). The transition arena process has also been used heuristically. Analysing a historical transition to improved stormwater quality treatment in Melbourne, Brown et al. (2013) reflect on the implications and lessons for transition management. One is that the main focus of transition management to date has been on the predevelopment phase of transitions with its focus on empowering frontrunners and niches (i.e., the transition arena process), whereas the acceleration phase of transitions might need a different focus and a better understanding of the institutional and policy context. Based on his work in a non-urban context—Dutch agriculture—Grin (2012) supports this conclusion regarding the role of frontrunners as helpful in accelerating developments but not sufficient; a larger group is needed to gain mass. More generally, not all scholars agree with a focus on selective participation of frontrunners, framing it as an 'elite group' (Smith and Stirling 2010), pointing to its legitimacy deficits (Hendriks 2009), and suggesting it as a problematic framing of an "*enlightened*"type of person (Jhagroe and van Steenbergen 2014, p. 2).

2.3.4 Transition Management Process Methodologies

Recent years have seen an adaptation of the framework and the instruments for the urban context in process methodologies or guidelines to be used either by (action) researchers (Wittmayer et al. 2011; Frantzeskaki et al. 2012a) or by local governments (Roorda et al. 2014) to implement a transition management approach in cities. In drawing up process methodologies for different 'target groups' (researchers, policy makers), different urban contexts (neighbourhoods, towns, cities), as well as different national contexts (different countries in Europe, Australia), the understanding of operational transition management has diversified (in terms of numbers of phases, levels of detail, attention to ethics, etc.). Although these process methodologies are far more specific and detailed in terms of process description than other transition management accounts, they still do not provide a clear-cut recipe: they need translation and adaptation to the specific transition challenges and questions in the urban context (Nevens et al. 2013; Wittmayer et al. 2014b).

By way of example, we turn to Roorda et al. (2014), who operationalised transition management into a process methodology for urban policy makers aiming for climate mitigation in their cities (Fig. 2.2). This specific process methodology has been developed in close collaboration between researchers and policy makers and was implemented in five European cities focussing on climate mitigation as part of the EU Interreg-funded MUSIC project (2010–2015) (see Wittmayer et al. 2016, Chap. 3, this volume). The process methodology distinguishes between different types of interventions that urban policymakers might use to influence the future of their city. It then outlines the different transition management instruments available for each of these more generic intervention types (see Fig. 2.2).

- Interventions aimed at *orienting* focus on positioning the city vis-à-vis societal developments and the municipality vis-à-vis other actors over time. Transition management instrumentsinclude, amongst others, system and actor analysis.
- Interventions aimed at *agenda-setting* focus on tactical governance activities in terms of integrating different agendas and practices and creating a sense of shared ownership and ambition for a sustainable future. Transition management instruments include, amongst others, transition agenda.
- *Activating* interventions focus on practices and setting up projects and experiments. Transition management instruments include transition experiments.
- Finally, interventions aimed at *reflecting* include the focus on supporting and enabling societal learning processes through both experience and cognitive engagement. Transition management instruments include transition experiments, monitoring, and evaluation.

The process methodology divides the intervention process into a number of phases, namely: (1) setting the scene for transition management, (2) exploring local dynamics, (3) framing the transition challenge, (4) envisioning a sustainable city, (5) reconnecting long term and short term, (6) engaging and anchoring, and

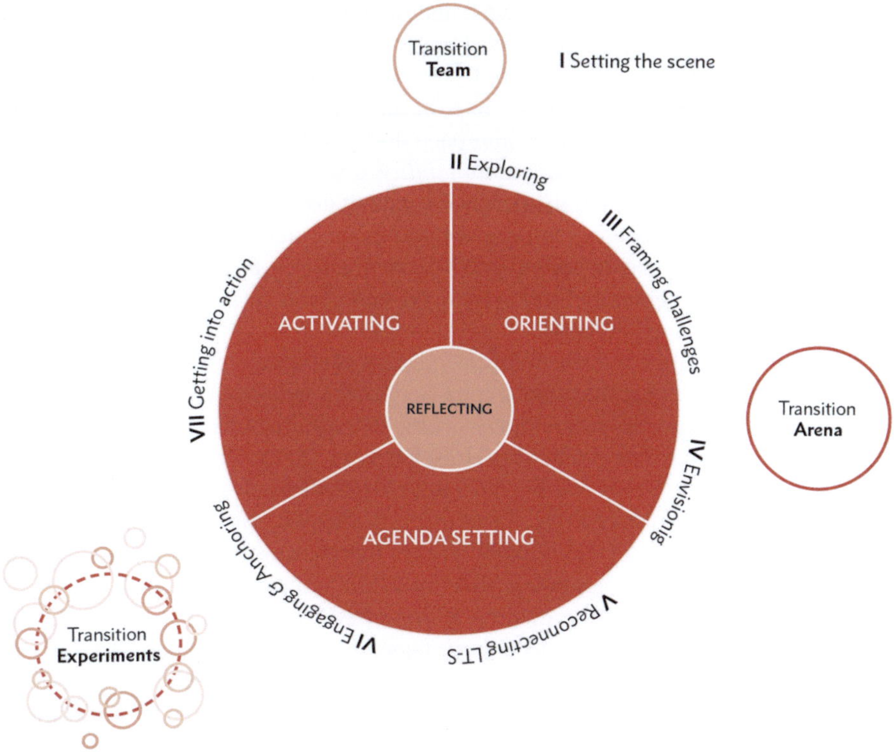

Fig. 2.2 The transition management process structure (from Roorda et al. 2014, p. 14)

(7) getting into action. These phases in turn are related to different settings or actors that foster interaction and focus on the emergence of alternative ideas, practices, and social relations; as such, it is an apt methodology for the predevelopment phase of transitions. The transition team, the transition arena, and the transition experiments (see Fig. 2.2) can be considered as actors and settings simultaneously. The transition team is a setting in which different individuals, such as urban policy makers, possibly specific actors from the city or transition experts, come together to negotiate the actual framing and embedding of the transition management instruments in the current (power and policy) context. As actor, the team is preparing and leading the actual transition management process. The transition arena simultaneously is the actor that is drawing up a new transition narrative and roadmap for the sustainable future of the city and the setting in which the urban frontrunners are negotiating this very future and agenda. In the same vein, the transition experiments are the actors that are practically addressing the societal challenges identified and consist of different frontrunners and stakeholders who experience the actual barriers and drivers for change by 'practising the transition.'

In the more operational applications of transition management, these process methodologies have been put into practice to organize contextualised transition management processes in cities, towns, and neighbourhoods (Nevens and Roorda 2014; Roorda and Wittmayer 2014; Wittmayer et al. 2013, 2014a, b; Ferguson et al. 2013; Frantzeskaki and Tefrati 2016, Chap. 4, this volume; Hölscher et al. 2016, Chap. 6, this volume; Krauz 2016, Chap. 8, this volume). Most of these accounts show that a transition management approach does not hold "*a silver bullet solution for actually realizing ambitious sustainability objectives*" (Nevens and Roorda 2014, p. 120). Nevertheless, transition management does provide an action impetus and more intangible outcomes in terms of practising collaborative governance and system thinking (Nevens and Roorda 2014), and it holds promises with regard to creating space for alternative ideas, practices, and social relationships (Wittmayer et al. 2014a; Roorda et al. 2014).

Many of the writings on these transdisciplinary operational processes witness the engagement of their authors with the earlier mentioned challenges of transition management in terms of the normative aim of sustainability (Wittmayer et al. 2014a), dis/empowerment dynamics (Hölscher et al. 2016, Chap. 6, this volume), the role of visioning (Frantzeskaki and Tefrati 2016, Chap. 4, this volume), or with regard to local power relationships (Krauz 2016, Chap. 8, this volume). Transition management processes in cities have shown that spaces for interaction can be created indeed, but that assuming that these are power-free spaces would be naïve. Especially when such a process is organised by a municipality, the risk is high that participants retreat to accustomed social roles and relations (Roorda and Wittmayer 2014). If a municipality usually relates to its citizens through public participation processes focussing on consultation, then a first step of a transition management-based process is to problematise the expectations towards one another. A necessary part of such a process is the experimentation with different expressions and meanings of social roles and relations (Wittmayer and van Steenbergen 2014; Wittmayer et al. 2014b). In this line, recent writings also show critical reflexivity in relationship to the roles of researchers in such processes (Wittmayer et al. 2014a; Wittmayer and Schäpke 2014).

Next to operational applications, we can see the process methodologies also being used as an analytical frame (i.e., heuristic application) to analyse existing governance dynamics. Shiroyama and Kajiki (2016, Chap. 7, this volume) use the operational framework by Roorda et al. (2014) to analyse the transition of the city Kitakyushu from an industrial to a green city by identifying transition arena, transition team, and transition experiment as settings and actors in this historical transition process.

2.4 Promises and Challenges of Governing Sustainability Transitions in Cities, Towns, and Neighbourhoods

Although applying transition management heuristically to cities and their governance does yield promising insights, such as with regard to the understanding of multi-actor governance processes, the nestedness of different geographic scales, and types of actors as well as the interrelatedness of developments in different domains, to date, most applications in the urban context have been operational applications of prescriptive process methodologies. In this section, we therefore first focus on synthesising the promises and challenges of transition management in cities for the more widely used operational applications (Sect. 2.4.1) before we focus on the characteristics of the urban context and its meaning for both heuristicand operational transition management processes (Sect. 2.4.2).

2.4.1 Promises and Challenges for Operational Applications of Transition Management

Transition management in the urban context is not a univocal success story, as outlined earlier. It is an approach in development. Considering that long-term transformation of any system "*will prove to be a messy, conflictual, and highly disjointed process*" (Meadowcroft 2009, p. 323), transition management in cities should not be considered a tool box or silver bullet, but rather an "*exploration of a new city governance approach for the co-creation of innovative pathways and processes in a stronglyreflexivemanner*" (Nevens et al. 2013, p. 121). Overall, challenges for operational transition management are related to the contextualisation of the approach to a specific societal challenge, actor constellation, place, and time; the fit with policy-making and decision-making institutions, as well as ongoing dynamics and developments; holding on to the radical character (i.e., directed at fundamental change); the importance of reflexivity and a space for learning, attention to politics and power relationships; and the degree to which sustainable development as the long-term normative goal can be made meaningful locally (see Nevens and Roorda 2014; Wittmayer et al. 2014a, 2015, 2016; Roorda and Wittmayer 2014).

Roorda et al. (2014) outline three promises of transition management in the context of urban climate governance; namely, it holds the potential to provide (1) a sense of direction for the city, (2) an impulse for local change, and (3) collective empowerment as it enables actors to address challenges and seize opportunities. Complementing ongoing regular policy processes and arenas as well as broader social movements and dynamics, operational applications of transition management create interactive spaces for *alternative ideas, practices, and social relations* in transdisciplinary settings (Wittmayer et al. 2014a), which have the potential to shift existing structures, cultures, and practices or 'transitionise' existing policies over time. In the following we use the distinction between impacts in terms of ideas, practices, and social relations to discuss the promises and challenges of operational applications of transition management.

Alternative *ideas* refer to a reframing of the actual challenges, alternative long-term directions, imaginations of the future, new discourses, and narratives through which actors involved gain a sense of urgency and the feeling that the impossible becomes possible. These new ideas and knowledge emerge through mutual and deep exchange, confrontation of opposing perspectives, and interaction of people from diverse backgrounds. Especially, the creation of alternative narratives can be seen as practising agency that opens up to the "*hypothetical, the possible, and the actual*" (Brockmeier 2009, p. 228). Through engaging in the creation of narratives and alternative futures, we "*undermine cultural norms and restrictions. It demonstrates that the mind interprets meanings as possibilities of action that reach beyond its own limits*" (ibid.). The challenge in engaging in a process of visioning or idea generation is the balance between opening up and fostering their plurality and diversity and closing down this process towards the convergence of a shared, albeit plural, notion of the future, for example, through the notion of a 'basket of future images' (cf. Stirling 2008).

In addition to probing what is possible through imagination, transition management is about creating space to *practising* alternatives—putting the imagination into action, done through projects, experimentation, and transformative action. There are manifold examples of best practises out there. The idea of experimentation is different: it is not about reading what others have done and copying it one-by-one, rather it is about defining a societal challenge and a way to address it through experimentation with a focus on learning by doing in a multi-actor setting. By engaging in action, actors learn about and find ways to address structural barriers as well as shape their future images (Van den Bosch 2010; Taanman et al. 2012).

In theory, no one actor is seen to be in the driving seat, or actually 'managing' a transition, which sets transition management aside, for example, from Local Agenda 21 processes, where more often than not the local government is in the lead and other actors in the urban society are invited to take part. In contrast, transition management aims to facilitate a joint societal searching and learning process in which ongoing actions by a range of actors are taken as a starting point to build new collaborative transition networks. As such, transition management opens a way to question and experiment with alternative *social relations*, such as between local governments and citizens, or between citizens and businesses. Policy institutions are both subject and object of transition governance: they can be important subjects in driving transition governance through their involvement and are also the object of transition as they are likely to change and gain a new understanding of their role and relationship to other actors. The emergence of new actors, such as the transition arena or follow-up networks, also questions and challenges the existing social fabric and local governance setting (Krauz 2016, Chap. 8, this volume); this immediately ties in with challenges and questions with regard to the kind of relations, the power, politics, norms, and ethics involved, as outlined earlier. Who is driving the process, with which agenda, and to what end? How does the process relate to incumbent actors? More often than not researchers have been involved in different capacities, which asks for reflexivity with regard to the different roles that a researcher might use in operational applications (Wittmayer and Schäpke 2014) and with regard to assumptions and frameworks used as well as specific ethical and scientific quality criteria.

2.4.2 *The Urban Context and Transition Management*

Referring back to the characteristics of the urban context outlined earlier (see Table 2.1), we discuss these here in terms of their meaning for operational and heuristic applications of transition management.

- *Geographic proximity*: In cities, the spatial distances between actors are usually shorter than, for example, in regions or nations. Actors in cities are physically closer to each other and share a certain geographically bounded area. As put by Boschma (2005, p. 59) "*Short distances bring people together, favour information contacts and facilitate the exchange of tacit knowledge*." For operational transition management processes, this means that being located in a city and being about a city (rather than about a 'national energy system') can increase identification with the area and create a shared purpose. There is also the risk of reifying administrative boundaries in delineating a system; for example, neighbourhood boundaries might not be recognised by actors (e.g., inhabitants) as such or might be an illogical confinement of inputs, activities, and impacts (cf. Wittmayer et al. 2013). Therefore, taking account of the construction of scale, to which we turn now, is important.
- *Multiscalar interaction*: Understanding cities as nested means that transition management applications, whether heuristic or operational, need to take multiscalar interactions into account. These scales can be national or international, neighbourhood or street, or any other geographic scale that is considered relevant. The city and 'its' actors actively construct relevant scales and interact with these in ways that support them in achieving their goals (cf. Coenen et al. 2012). Through transition governance applications we can analyse this interaction as a two-way street and as such play into it. Cities may, for example, refer to EU-level strategies (e.g., Europe 2020) or EU-wide covenants (e.g., Covenant of Mayors), to further their own ambition of CO_2 reduction, bypassing national governance. Through their construction and interpretation of and reaction to certain events (such as budget cuts) cities can be inspiring other cities but also initiate new legislation on the national or international level.
- *Multi-domain interaction*: Taking a place-based system delineation involves that transition governance activities are not only taking account of changes in one domain, rather it is in actual places where changes in different domains (energy, mobility, water, . . .) come together and interact. As such, a place-based approach to transitions involves the multitude of dynamics between different domains in a specific place, increasing the complexity of the task at hand, but also providing numerous points of leverage. Working on CO_2 reductions means that the process will focus not only on issues of energy provision and production but rather, in the process of problem framing and future visioning, have a broad and integral perspective that also encompasses issues in domains such as mobility, water, lifestyle, and tourism.

- *Personal proximity*: relates to the concept of social proximity (Boschma 2005): Cities, towns, and neighbourhoods are also environments in which people live, love, rage, or die. It is people in their roles as *inhabitants, fathers, mothers, or engaged neighbours* who become actors in transition governance activities, rather than (only) as *professionals* as is the case in many transition management processes in functional systems. People are involved in different roles and have clear personal, emotional, and social stakes as well as trust relationships: they live in the city, raise their children there, or cheer for the local football club—all these relationships are embedded and come with certain expectations and responsibilities. This definition makes urban transition management a collective endeavour of people striving for sustainable development in their own living environment and brings powerstruggles and the search for new roles and relations very close to the individual and his or her homestead.
- *Institutional proximity*: refers to proximity that originates in shared formal and informal institutions including laws and rules as well as cultural norms and habits (Boschma 2005). For certain issues, there might be a high extent of institutional proximity within a city (e.g., formalised governance processes), whereas for other issues this might be lower (e.g., if the city's population is composed of people from different national or cultural backgrounds). Transition management activities aim at changing institutional structures, cultures, and practices (Frantzeskaki et al. 2012b), and as such are working on creating new institutional proximity. In doing so, they work at the fringes of existing institutions (cf. Coenen et al. 2012).

2.5 Conclusion

The transition management-based analysis and interventions over the past years, including those described in this volume, have led to a more systemic, contextual, and effective way to develop alternative ideas, practices, and social relations. As a counterbalance to optimisation of existing systems, transition management thus aids in strengthening alternative dynamics and empowering actors to seek to change existing unsustainable systems. In the light of the changing contexts and dynamics and as actual transitions accelerate, it is increasingly evident that new and additional governance mechanisms need to be developed (Loorbach 2014). In contexts where the need or desirability of transitions is no longer an issue, alternatives are rapidly diffusing and incumbent regimes are fragmenting, adapting, and eroding. This pivotal point is where new forms of top-down and formal policy are needed to help institutionalize new rules that emerge, as well as to stop investment in and work on unsustainable development. Especially, this latter point relates to the necessity of breaking down barriers and unsustainable practices in a more or less systematic way. As local renewable energy production becomes superior to centralised fossil fuel-based energy, policy at a certain point needs to phase out (its dependence on tax income from) fossil energy, creating a new norm—which then puts power issues centre stage. A challenge for the coming decade, it seems now, is to understand, analyse, and create breakthroughs in existing power structures by interlinking change-inclined regime members to emergent new power structures, next to developing alternatives and countermovements.

References

Avelino F (2009) Empowerment and the challenge of applying transition management to ongoing projects. Policy Sci 42(4):369–390

Avelino F, Loorbach D, Wittmayer J (2011) Het Stimuleren van Duurzaamheid in de Stad. Een Onderzoeksadvies aan de Gemeente Den Haag [commissioned by Gemeente Den Haag]. DRIFT, Rotterdam

Avelino F, Bressers N, Kemp R (2012) Transition management as new policy making for sustainable mobility. In: Geerlings H, Shiftan Y, Stead D (eds) Transition towards sustainable mobility: the role of instruments, individuals and institutions, pp 33–52. Ashgate, Hampshire

Boschma RA (2005) Proximity and innovation: a critical assessment. Reg Stud 39(1):61–74

Brockmeier J (2009) Reaching for meaning: human agency and the narrative imagination. Theor Psychol 19(2):213–233

Brown RR, Farelly MA, Loorbach D (2013) Actors working the institutions in sustainability transitions: the case of Melbourne's stormwater management. Glob Environ Change 23(4):701–718

Coenen L, Truffer B (2012) Places and spaces of sustainability transitions: geographical contributions to an emerging research and policy field. Eur Plan Stud 20(3):367–374

Coenen L, Benneworth P, Truffer B (2012) Toward a spatial perspective on sustainability transitions. Res Policy 41:968–979

Ferguson B, Frantzeskaki N, Brown R (2013) A strategic program for transitioning to a water sensitive city. Landsc Urban Plan 117:32–45

Frantzeskaki N, Tefrati N (2016) A transformative vision unlocks the innovative potential of Aberdeen City, UK. In: Loorbach D, Wittmayer J, Shiroyama H, Fujino J, Mizuguchi S (eds) Governance of urban sustainability transitions, pp 53–72. Springer, Tokyo

Frantzeskaki N, Ferguson BC, Skinner R, Brown RR (2012a) Guidance manual: key steps for implementing a strategic planning process for transformative change. Dutch Research Institute for Transitions, Erasmus University Rotterdam. Monash Water for Liveability, Monash University, Melbourne. ISBN 978-1-921912-14-6

Frantzeskaki N, Loorbach D, Meadowcroft J (2012b) Governing transitions to sustainability: transition management as a governance approach towards pursuing sustainability. Int J Sustain Dev 15(1/2):19–36

Frantzeskaki N, Wittmayer J, Loorbach D (2014a) The role of partnerships in 'realizing' urban sustainability in Rotterdam's City Ports Area, the Netherlands. J Clean Prod 65:406–417

Frantzeskaki N, Bach M, Hölscher K, Wittmayer JM, Loorbach D (2014b) Ten years of transition management: taking stock and looking forward. Presented at the 5th international conference on sustainability transitions: impacts and institutions, Utrecht, 27–29 Aug 2014

Grin J (2012) The politics of transition governance in Dutch agriculture. Conceptual understanding and implications for transition management. Int J Sustain Dev 15(1/2):72–89

Grin J, Rotmans J, Schot J, i.c.w. Geels F, Loorbach D (2010) Transitions to sustainable development – part 1. New directions in the study of long term transformative change. Routledge, New York

Hendriks C (2009) Policy design without democracy? Making democratic sense of TM. Policy Sci 42(4):341–368

Hölscher K, Roorda C, Nevens F (2016) Ghent – fostering a climate for transition. In: Loorbach D, Wittmayer J, Shiroyama H, Fujino J, Mizuguchi S (eds) Governance of urban sustainability transitions, pp 95–115. Springer, Tokyo

ICLEI (2012) Local sustainability 2012: taking stock and moving forward, Global review, ICLEI global report. ICLEI, Freiburg

Jhagroe S, Loorbach D (2014) See no evil, hear no evil: the democratic potential of transition management. Environ Innov Soc Transit 15:65–83

Jhagroe S, van Steenbergen F (2014) Van koploper naar koplopen: een pleidooi voor inclusief leiderschap. Drift essay 2014.04. Online at: http://www.drift.eur.nl/?p=262. Jan 2015

Kemp R, Rotmans J (2009) Transitioning policy: co-production of a new strategic framework for energy innovation policy in the Netherlands. Policy Sci 42:303–322

Kemp R, Schot J, Hoogma R (1998) Regime shifts to sustainability through processes of niche formation: the approach of strategic niche management. Tech Anal Strat Manag 10(2):175–196

Kern F (2012) The discursive politics of governing transitions towards sustainability: the UK Carbon Trust. Int J Sustain Dev 15(1/2):90–105

Kern F, Howlett M (2009) Implementing transition management as policy reforms: a case study of the Dutch energy sector. Policy Sci 42(4):391–408

Krauz A (2016) Transition management in Montreuil: towards perspectives of hybridization between 'top-down' and 'bottom-up' transitions. In: Loorbach D, Wittmayer J, Shiroyama H, Fujino J, Mizuguchi S (eds) Governance of urban sustainability transitions, pp 137–154. Springer, Tokyo

Lang DJ, Wiek A, Bergmann M, Stauffacher M, Martens P, Moll P, Swilling M, Thomas C (2012) Transdisciplinary research in sustainability science – practice, principles and challenges. Sustain Sci 7(suppl 1):25–43

Leach M, Scoones I, and Stirling A (2010) Dynamic Sustainabilities: technology, environment, social justice. Earthscan, Oxon

Loorbach D (2007) Transition management. New mode of governance for sustainable development. PhD thesis, Erasmus University, Rotterdam

Loorbach D (2010) Transition management for sustainable development: a prescriptive, complexity-based governance framework. Governance 23(1):161–183

Loorbach D (2014) To transition! Governance panarchy in the new transformation. DRIFT/EUR, Rotterdam

Loorbach D, Rotmans J (2012) Transities en transitiemanagement. Oorsprong, status en toekomst. E2012.01. DRIFT, Rotterdam

Loorbach D, Shiroyama H (2016) The challenge of sustainable urban development and transforming cities. In: Loorbach D, Wittmayer J, Shiroyama H, Fujino J, Mizuguchi S (eds) Governance of urban sustainability transitions, pp 3–15. Springer, Tokyo

Meadowcroft J (2009) What about the politics? Sustainable development, transition management, and long term energy transitions. Policy Sci 42(4):323–340

Mizuguchi S, Ohta K, Beer PJ, Yamaguchi M, Nishimura T (2016) Interactions among multiple - niche-innovations and multi-regimes: the case of the "welfare mall" in Higashiomi. In: Loorbach D, Wittmayer J, Shiroyama H, Fujino J, Mizuguchi S (eds) Governance of urban sustainability transitions, pp 73–93. Springer, Tokyo

Nevens F, Roorda C (2014) A climate of change: a transition approach for climate neutrality in the city of Ghent (Belgium). Sustain Cities Soc 10:112–121

Nevens F, Frantzeskaki N, Loorbach D, Gorissen L (2013) Urban transition labs: co-creating transformative action for sustainable cities. J Clean Prod 50:111–122

Olsson P, Gunderson LH, Carpenter SR, Ryan P, Lebel L, Folke C, Holling CS (2006) Shooting the rapids: navigating transitions to adaptive governance of social-ecological systems'. Ecol Soc 11(1):18

Quist J, Thissen W, Vergragt P (2011) The impact and spin-off of participatory backcasting after 10 years: from vision to niche. Technol Forecast Soc Chang 78(5):883–897

Quist J, Wittmayer JM, van Steenbergen F, Loorbach D (2013) Combining backcasting and transition management in the community arena. In: Quist J, Wittmayer J, Umpfenbach K, Bauler

T (eds) Pathways, transitions and backcasting for low-carbon and sustainable lifestyles. Sustainable consumption transitions series, issue 3. Proceedings of SCORAI Europe & InContext workshop, 7–8 Oct 2013, Rotterdam. Online available at: http://www.incontext-fp7.eu/sites/default/files/Proceedings%20InContext%20SCORAI%20Quist%20etal%20p33-54.pdf

Raven R, Schot J, Berkhout F (2012) Space and scale in socio-technical transitions. Environ Innov Soc Transit 4:63–78

Roorda C, Wittmayer J (2014) Transition management in five European cities – an evaluation. DRIFT, Erasmus University Rotterdam, Rotterdam

Roorda C, Wittmayer J, Henneman P, van Steenbergen F, Frantzeskaki N, Loorbach D (2014) Transition management in the urban context. Guidance manual. DRIFT, Erasmus University Rotterdam, Rotterdam

Rotmans J, Loorbach D (2009) Complexity and transition management. J Ind Ecol 13(2):184–196

Rotmans J, Kemp R, Van Asselt M (2001) More evolution than revolution: transition management in public policy. Foresight 3(1):15–31

Schot J, Geels FW (2008) Strategic niche management and sustainable innovation journeys: theory, findings, research agenda, and policy. Tech Anal Strateg 20(5):537–554

Shiroyama H, Kajiki S (2016) Case study of eco-town project in Kitakyushu: tension among incumbents and the transition from industrial city to green city. In: Loorbach D, Wittmayer J, Shiroyama H, Fujino J, Mizuguchi S (eds) Governance of urban sustainability transitions, pp 117–136. Springer, Tokyo

Shove E, Walker G (2007) CAUTION! Transitions ahead: politics, practice, and sustainable transition management. Environ Plan A 39:763–770

Smith A, Stirling A (2010) The politics of social-ecological resilience and sustainable socio-technical transitions. Ecol Soc 15(1):11

Smith A, Stirling A, Berkhout F (2005) The governance of sustainable socio-technical transitions. Res Policy 34:1491–1510

Sondeijker S (2009) Imagining sustainability: methodological building blocks for transition scenarios. PhD thesis, Erasmus University Rotterdam, Rotterdam

Stirling A (2008) Opening up and closing down: power, participation and pluralism in the social appraisal of technology. Sci Technol Hum Values 33(2):262–294

Taanman M (2014) Looking for transitions. Monitoring approach for sustainable transition programmes. PhD thesis, Erasmus University Rotterdam, Rotterdam

Taanman M, Wittmayer JM, Diepenmaat H (2012) Monitoring on-going vision development in system change programmes. J Chains Netw Sci 12(2):125–136

United Nations Conference on Environment and Development (1992) Agenda 21. UNCED, New York. Available from: http://sustainabledevelopment.un.org/content/documents/Agenda21.pdf. Accessed 18 Feb 2014

Van Buuren A, Loorbach D (2009) Policy innovation in isolation. Conditions for policy-renewal by transition arenas and pilot projects. Public Manag Rev 11:375–392

Van den Bosch S (2010) Transition experiments: exploring societal changes towards sustainability. Doctoral dissertation, Erasmus University Rotterdam, Rotterdam

Van den Bosch S, Rotmans J (2008) Deepening, broadening and scaling up. A framework for steering transition experiments. KCT essay #02. Knowledge Centre for Sustainable System Innovations and Transitions (KCT), Delft and Rotterdam

Van der Brugge R, Rotmans J, Loorbach D (2005) The transition in Dutch water management. Reg Environ Chang 5:167–176

Verbong G, Loorbach D (eds) (2012) Governing the Energy Transition. Reality, Illusion or Necessity? Routledge, New York

Voß JP (2014) Performative policy studies: realizing "transition management". Innov Eur J Soc Sci Res 27(4):317–343

Voß JP, Bauknecht D, Kemp R (2006) Reflexive governance for sustainable development. Edward Elgar, Cheltenham/Northampton

Wiek A (2007) Challenges of transdisciplinary research as interactive knowledge generation – experiences from transdisciplinary case study research. GAIA 16(1):52–57

Wiek A, Talwar S, O'Shea M, Robinson J (2014) Towards a methodological scheme for capturing societal effects of participatory sustainability research. Res Eval 23(2):117–132

Wittmayer J (2014) Creating a sustainability movement in The Hague, The Netherlands. In: Wittmayer J, Roorda C, van Steenbergen F (eds) Governing urban sustainability transitions – inspiring examples. DRIFT, Rotterdam, pp 68–73

Wittmayer J (2016) Insights and lessons for the governance of urban sustainability transitions. In: Loorbach D, Wittmayer J, Shiroyama H, Fujino J, Mizuguchi S (eds) Governance of urban sustainability transitions, pp 157–173. Springer, Tokyo

Wittmayer JM, Schäpke N (2014) Action, research and participation: roles of researchers in sustainability transitions. Sustain Sci 9(4):483–496. doi:10.1007/s11625-014-0258-4

Wittmayer J, van Steenbergen F (2014) Researching roles in transition: a case study. Paper presented at the 9th international conference in interpretive policy analysis. Governance and beyond: knowledge, technology and communication in a globalizing world, Wageningen, 3–5 July, 2013

Wittmayer J, van Steenbergen F, Quist J, Loorbach D, Hoogland C (2011) The community arena: a co-creation tool for sustainable behaviour by local communities. Methodological guidelines. Deliverable 4.1, InContext: EU ENV.2010.4.2.3-1 grant agreement n° 265191. Online available here: http://www.incontext-fp7.eu/sites/default/files/D4-1_Methodological%20guidelines_final.pdf

Wittmayer J, van Steenbergen F, Baasch S, Feiner G, Mock M, Omann I (2013) Pilot projects rounding up. Year 3 pilot-specific report. Deliverable 4.4. InContext: EU ENV.2010.4.2.3-1 grant agreement n° 265191. Online available here: http://www.incontext-fp7.eu/sites/default/files/Pilot%20projects%20rounding%20up%2C%20Year%203%20Pilot-Specific%20Report.pdf

Wittmayer JM, Schäpke N, van Steenbergen F, Omann I (2014a) Making sense of sustainability transitions locally. How action research contributes to addressing societal challenges. Crit Policy Stud 8(4):465–485

Wittmayer J, Roorda C, van Steenbergen F (eds) (2014b) Governing urban sustainability transitions: inspiring examples. DRIFT, Rotterdam

Wittmayer JM, Rok A, Roorda C, van Steenbergen F (2015) Governing sustainability: a dialogue between Local Agenda 21 and transition management. Local Environ. doi:10.1080/13549839.2015.1050658

Wittmayer J, Mizuguchi S, Rach S, Fujino J (2016) City networks for sustainability transitions in Europe and Japan. In: Loorbach D, Wittmayer J, Shiroyama H, Fujino J, Mizuguchi S (eds) Governance of urban sustainability transitions, pp 37–50. Springer, Tokyo

Chapter 3
City Networks for Sustainability Transitions in Europe and Japan

Julia M. Wittmayer, Satoru Mizuguchi, Sarah Rach, and Junichi Fujino

Abstract In both Europe and Japan, city networks have emerged aiming at supporting cities to collectively address sustainability challenges. In this chapter, we introduce four such city networks: the *MUSIC network* in Europe and the *Eco-model City*, the "Future City" Initiative, and the *Green and local autonomy model city* in Japan. The numerous sustainability challenges of the networked cities include the reduction of CO_2 emissions, the integration of sustainability in urban planning processes, or demographic change. Notwithstanding similar aims, these networks have a different setup and thereby illustrate different possibilities to address sustainability challenges on an urban scale through concerted action by a network of cities. The *MUSIC network* is a cooperation project among European cities and research institutes in Northwest Europe, with the aim to reduce CO_2 emissions by 50 % by 2030 in the five European partner cities: Aberdeen, Montreuil, Ghent, Ludwigsburg, and Rotterdam. The *Eco-model Cities*, "Future City" Initiative, and *Green and local autonomy model city* networks focus on innovating technologies, services, and business models as well as the socioeconomic and physical fabric of the cities to make them fit for addressing global developments. All four city networks systematically organise learning and exchange between the cities on their pathways to becoming more sustainable.

Keywords City networks • Learning • Low-carbon society • Sustainable development • City collaboration

J.M. Wittmayer (✉) • S. Rach
Dutch Research Institute for Transitions, Erasmus University Rotterdam, Rotterdam, The Netherlands
e-mail: wittmayer@drift.eur.nl

S. Mizuguchi
Japanese Forum of Environmental Journalists, Tokyo, Japan

J. Fujino
National Institute of Environmental Studies, Tokyo, Japan

D. Loorbach et al. (eds.), *Governance of Urban Sustainability Transitions*, Theory and Practice of Urban Sustainability Transitions,
DOI 10.1007/978-4-431-55426-4_3

3.1 City Networks for Sustainable Development

Cities all across the world have long been in fruitful exchange with each other, especially through trade relations. At the beginning of the twentieth century, another kind of relationship emerged, one based on a common aim for a better world. Examples are city twinning aimed at enhancing mutual understanding and alliances after the Second World War (Zelinsky 1991) or cities linking up to face sustainability challenges together (ICLEI 2012). This twinning and linking has been supported by national and supranational organisations, such as the Council of European Municipalities and Regions, the International Union of Local Authorities, Local Governments for Sustainability, the European Sustainable Cities and Towns campaign, or the United Nations Environment Programme (Zelinsky 1991; ICLEI 2012). As such, national and supranational organisations as well as associations have long been a part of creating ties between cities and municipalities to support the creation of a more sustainable world.

In this chapter, we introduce four city networks in Europe and Japan that aim to support cities in working towards a sustainable future from a sustainability transitions perspective: the *MUSIC* network in Europe and the *Eco-model city*, the *"Future City" Initiative (FCI)*, and the *Green and local autonomy model city* in Japan. We chose these networks for being inspired by or applying transition thinking in urban areas. The aim of the networks is to support cities in addressing their sustainability challenges. These challenges are numerous and include the reduction of CO_2 emissions, the integration of sustainability in urban planning processes, or demographic change. Notwithstanding similar aims, these networks have different setups and thereby illustrate different possibilities to address sustainability challenges on an urban scale through concerted action by a network of cities (Table 3.1).

The networks support partner cities in addressing sustainability challenges primarily by offering space for formal and informal learning. This learning includes gaining insights into other and new perspectives, exchanging formal and informal practices within administrations, discovering, experimenting with, and experiencing new approaches, methods, and tools through which to address sustainability challenges. It also leads to a further development of approaches and tools, for example, the GIS tools and the transition management approach for the urban context within the MUSIC network.

The chapter is based on the experiences of the authors as well as on additional reading of the primary and secondary literature on these networks. The authors were involved in these networks as researcher (first author), academic advisor at the selection committee for the FCI (fourth author), and unofficial participant for 'visioning' meetings of FCI (second author). This chapter introduces the networks outlined in Table 3.1, including their funding context, rationale, and role in supporting cities in their sustainability ambitions. In the next section, we first introduce the European network (Sect. 3.2), followed by the Japanese networks (Sect. 3.3). The final Sect. 3.4 introduces the five cities that are part of either of the

Table 3.1 Overview of city networks

	MUSIC network	Eco-model cities	"Future City" Initiative	Green and local autonomy model city
Initiated by	European Union	Government of Japan	Government of Japan	Government of Japan
Funded by	European Union and network partners	Government of Japan and network partners	Government of Japan and network partners	Government of Japan and network partners
Funding amount	€ 2,810,714 (European Regional Development Fund contribution, plus own contribution by cities and research institutes)	No specific budget of Cabinet Office, but governmental agencies funded some projects (amount not disclosed)	$10,000,000 (Cabinet Office budget), plus own contribution by cities, citizens and business	$30,000 per model city (budget of Ministry of General Affairs), plus own contribution by cities, citizens and business
Funding period	2009–2014, extension for one year approved in November 2013	2008–2012; with an extra addition of cities in 2012–2013	2012–2013 (governmental funding period)	2011 (governmental funding period)
Partners	*Cities*: Aberdeen (UK), Ghent (BE), Ludwigsburg (D), Montreuil (F), Rotterdam (NL) *Research institutes*: DRIFT Erasmus University Rotterdam (NL), Henri Tudor (LU)	Shimokawa, Obihiro (Hokkaido, Northern Island), Chiyoda, Yokohama, Iida, Toyama, Toyota, Kyoto, Sakai (main island), Yusuhara (Shikoku Island), Kitakyushu, Minamata (Kyushu Island), Miyakojima (Southern small island)	Shimokawa, Kamaishi[a], Kesen regional area[a], Iwanuma[a],Shinchi[a], Higashimatsushima[a], Minamisoma[a] Toyama, Kitakyushu, Yokohama, Kashiwa (main island)	Hachinohe, Tono (North-East region of main island), Higashiohmi (main island)
Geographical spread	North-West Europe	Whole of Japan	Whole of Japan	Main island of Japan
Exclusiveness	The network is closed in that it is not possible for other cities to join, after the funding had been approved by the European Union	13 cities were selected in a first round, ±40 cities were to be selected by 2016	11 cities were selected, the number will be about 20 cities by 2016	The network is closed in that it is not possible for other cities to join
Ambition/aim	Focus on energy challenge: the reduction of CO_2 emissions by 50 % until 2030	Demonstrating the image of a low-carbon society	Addressing a variety of challenges like the environment, super ageing and green growth. Followed by international dissemination	Demonstrating the image of local autonomy with green growth
Activities	• Create space for exchange and learning • Develop transition management for the urban context • Develop GIS tools and modules for use for urban energy planning • Implement transition management in partner cities • Implement GIS tools and modules in partner cities	• Aim for integrated urban planning, • Create space for exchange and learning • Develop action plan with assessment methods and indexes • Publish good practices	• Aim for integrated urban planning, • Create space for exchange and learning • Disseminate internationally, through participating in conferences and organizing conferences	• Organize best practice of local autonomy and green development • Create space for exchange and learning
Role of science/transition thinking	Transition thinking was explicitly applied in the project. Transition management was used as a prescriptive methodology to be applied in the five partner cities	Transition thinking was not explicitly applied, but some cities had organized transition arena without knowing the theory	Transition thinking was introduced by transition researchers at an international conference and began to influence the whole project	Transition thinking was not explicitly applied, but some cities had organized transition arena without knowing the theory

[a]Suffered from earthquake

networks and which are featured as case studies in the chapters of this book: Aberdeen (Frantzeskaki and Tefrati 2016, Chap. 4, this volume), Higashiohmi (Mizuguchi et al. 2016, Chap. 5, this volume), Ghent (Hölscher et al. 2016, Chap. 6, this volume), Kitakyushu (Shiroyama and Kajiki 2016, Chap. 7, this volume), and Montreuil (Krauz 2016, Chap. 8, this volume).

3.2 The MUSIC Network

One of the instruments of the European Cohesion and Regional Policy is the Interreg-funding scheme of the European Union that is promoting and funding cooperation of regions, cities, and networks.[1] It distinguishes between different phases of funding and different types of cooperation (cross-border, transnational, and interregional). From 2007 to 2013 (the fourth phase), transnational cooperation was endorsed among different areas, one being the North-West Europe region. A total of 355 million euros of the European Regional Development Fund (ERDF) were invested to "*promote the economic, environmental, social and territorial future of the North-West Europe area*" (Interreg IVB NWE 2013a) and as such contribute to the goals of EU2020, the 10-year strategy to strengthen the European Union (EU) economy.

According to this strategy, economic growth in the EU should be (amongst other concerns) sustainable through furthering a low-carbon economy. One of the funded projects in the new region that is addressing this aim is the *MUSIC* project, short for *Mitigation in Urban Areas: Solutions for Innovative Cities* (Interreg IVB NWE 2013b). It is a transnational cooperation project among five European cities (Aberdeen, Montreuil, Ghent, Ludwigsburg, and Rotterdam), which are supported by two research institutes, the Dutch Research Institute for Transitions (DRIFT, The Netherlands) and the Public Research Centre Henri Tudor (Luxembourg) (Fig. 3.1). There was no external selection process; rather, the participating cities and institutes jointly wrote a funding proposal at the outset. The project runs from 2010 to 2015 and has an overall budget of 2.8 million euros. Half this budget comes from the Interreg funding scheme; the other half is co-financed by the cities and research institutes. The money was spread across three types of activities, as outlined next, with a major part being spend on the implementation of a pilot project.

Inspired by the EU's aim to decrease CO_2 emissions by 20 % by 2020, this network of cities and research institutes formulated the goal to reduce CO_2 emissions to 50 % by 2030 in the five cities, as such living up to the commitment they

[1] There are also other funding schemes such as URBACT, also an instrument of European Cohesion Policy and co-financed by the European Regional Development Fund, the 28 Member States, Norway, and Switzerland. It aims to foster sustainable integrated urban development in cities across Europe.

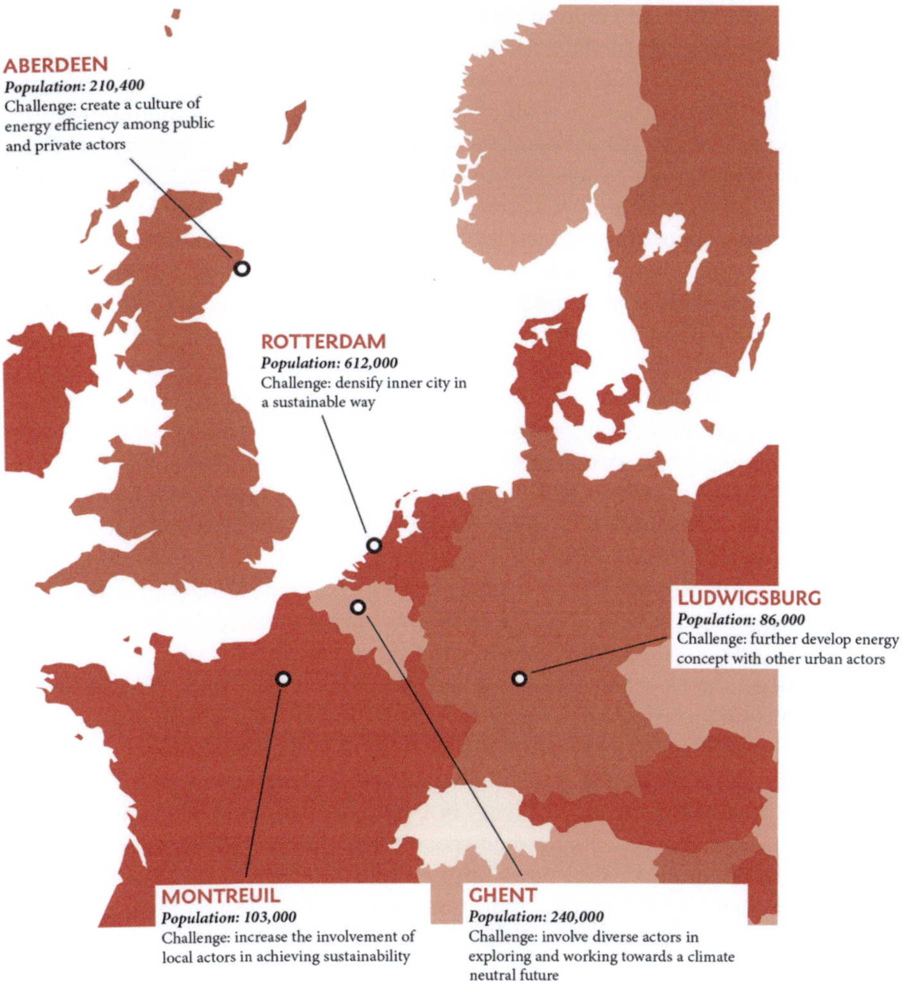

Fig. 3.1 Overview of the five cities of the *MUSIC* project (Source: Roorda et al. 2014, p. 8)

gave when signing the Covenant of Mayors[2] and their cities' ambitions for sustainability.

To reach these ambitions, the cities aim to make CO_2 reduction an integral part of their urban planning processes. The MUSIC project supports them in three ways in making this transition. First, the cities implement a transition management approach, through which a local sustainability vision and action plan are developed by reaching out to urban change agents from different backgrounds. Second, a new Geospatial Information System (GIS) tool is developed that allows for smart urban

[2] Aberdeen, Ghent, Montreuil, and Rotterdam signed in 2009, Ludwigsburg in 2012.

energy planning through offering different modules (e.g., solar, energy poverty, heat islands) to ease identification of possibilities and challenges for CO_2 reduction and renewable energy production in all the cities. These two activities focus on innovating local cooperation as well as the use of data and are supported by, third, the implementation of a pilot project, an actual measure for reducing CO_2 emission in public buildings, such as schools or community centres.

Not only within the cities, but also within the network and between the cities, the MUSIC network opened a space for learning from and with each other (cf. Roorda 2013). Part of this learning was organised and coordinated by the project lead committee, such as the transnational network meetings. Learning took also place within each individual city context through the coaching of city officers with regard to the implementation of transition management as well as the use and development of the GIS tools. In addition, a mid-term and a final conference as well as transition management workshops were organised, wherein outcomes were shared and additional cities could familiarise themselves with the approaches developed.

The MUSIC cities met in half-yearly transnational network meetings between 2010 and 2015. These meetings were mainly aimed at transnational exchange of experiences with the transition management approach as well as the development of the GIS tools (cf. Roorda 2013). These meetings harboured training sessions on both transition management and GIS, and gave ample opportunity to the cities to present their own experiences and learn from the experiences of others. They were also designed to put specific problems and questions on the table through which both the transition management approach, which to date had not been used in cities as well as the GIS tools, could be further developed and contextualised.

Each city had a dedicated transition coach from DRIFT (Erasmus University Rotterdam) to support the city officers in contextualising and implementing the transition management process for their specific city. These transition coaches have been visiting the cities several times with an intensive period of coaching in the very beginning of implementing transition management. Rather than a one-off implementation of a participatory process, the use of transition management was to a great extent about learning new ways to perceive the city and its challenges as well as to act upon this new perspective; for example, accepting ambiguity, uncertainty, and complexity and taking these characteristics as opportunity rather than as constraints in governing a city. As such it was a learning process for finding more integrated ways to sustainable urban planning. A similar approach was taken for developing GIS tools and components, which were to be customised for each city in teams of city officers and GIS experts from the Henri Tudor Institute. For example, one city would focus on energy poverty and needed tools to analyse and visualise this, and another would focus on identifying ‘heat islands’.

Beside the half-yearly transnational workshops, each city organised transition arenas. Being a central instrument in transition management, a transition arena is a setting that provides an informal, well-structured space to a small group of change agents with diverse perspectives (see Wittmayer and Loorbach 2016, Chap. 2, this volume). In this setting, the change agents structure the transition challenge of their city, envision a sustainable future, and develop transition pathways as well as actual transition experiments. The ideas, connections, and energy that emerged during the process have proven to be a fertile ground for furthering strategies, initiatives, and collaborations aimed at a sustainable future of the city. The exchange of the experiences in this process between different cities proved to be highly relevant and inspiring.

3.3 Three City Networks for a Low-Carbon Society: Japan

Towards the end of the first decade of the 2000s, people's awareness of climate change was high. In 2007 the Intergovernmental Panel for Climate Change (IPCC) had received the Nobel Peace prize, and a year later Japan presided over the G8 Summit with climate change as one of the central topics. The first commitment period of the Kyoto Protocol on Climate Change was between 2008 and 2012, and in 2009, the 15th Conference of the Parties on Climate Change (COP15) took place, with high international expectations with regard to a new climate agreement (United Nations Institute for Training and Research 2015). It was against this background that the Japanese Government initiated three city development programs to promote a low-carbon society. We outline these in the following sections.

3.3.1 *Eco-model Cities*

The *Eco-model Cities* program, started in 2008 and aimed at demonstrating the image of a low-carbon society which Japan aspired to be in the future (Promotion Council for the "Future City" Initiative 2014). The same year 13 cities were selected (Fig. 3.2) based on their ambitious reduction goals as well as past performance in this regard.

After the selection, each city submitted a specific 'Action Plan' that included its own assessment methods, evaluation indexes, and goals. A governmental body, composed of the secretariat of the Cabinet Office, an inter-ministerial coordinating body, and an academic advisory group, examined the plans. Before disclosure of the plans to the public, the governmental body made suggestions and gave advice to increase the effectiveness of the plans. The good practices of the selected cities were published to allow other municipalities to take over specific elements. The Eco-model Cities and other motivated municipalities and related organisations participated in the establishment of the Promotion Council for Low-carbon Cities in December 2008. This Council promoted learning from the trial and errors of the selected Eco-model Cities (Cabinet Office 2013).

Although there was next to no designated budget for the selected cities, they could profit from easier access to related ministries for tapping into other appropriate subsidies (Murakami 2012) and from the mobilisation effect that the labelling

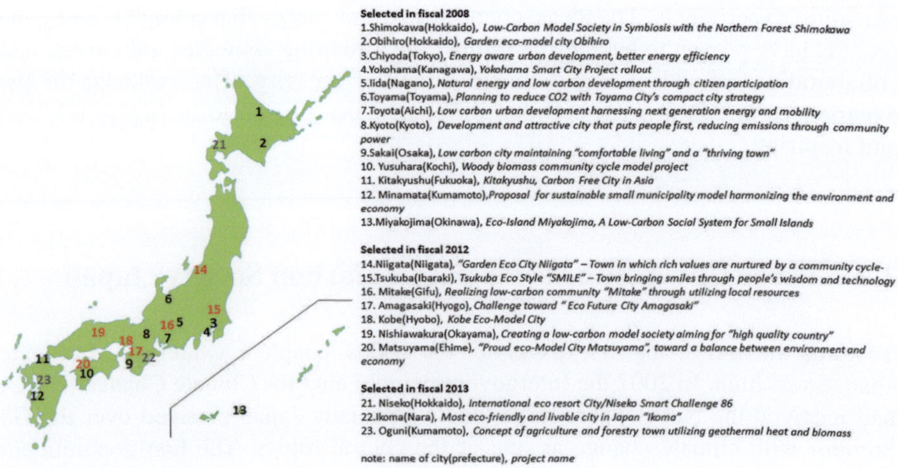

Fig. 3.2 Location map of *Eco-model Cities*

as Eco-model Cities had on citizens and business to engage in climate action and sustainable urban development.

In 2012 and 2013, additional cities were selected (Fig. 3.2), adding up to a total of 23 cities at present. By 2016, the number of selected cities was to be about 40–50 to cover almost all prefectures (Cabinet Office 2013). In 2014, the initiative was merged with the "Future City" Initiative (FCI), to which we turn now.

3.3.2 "Future City" Initiative

After the general elections in 2010, a new government was formed by the Democratic Party of Japan. Against the background of an ageing rate of 24 % in 2009, which is expected to rise to 40 % by 2030, as well as an economy that has been stagnating for two decades, this government formulated a 'New Growth Strategy' to boost the economy through green innovation. The FCI constituted part of the strategy aiming at reducing greenhouse gas, tackling the super-ageing society, and boosting the local economy. It was launched in 2011.

In 2014, 11 municipalities were selected (Fig. 3.3) by a committee composed of bureaucrats and experts from climate science, energy, industrial design, engineering, urban development, finance, and governance. Excellency of the long-term vision as well as the potentiality of the proposed consortium composed by academics, business, financial institutions, etc. were key criteria to become a "Future City" (Cabinet Office 2008). Among the 11 selected municipalities, 6 were from the areas that were hit by the great earthquake of 2011; the remainder had been Eco-model Cities.

Fig. 3.3 Location map of the *Future City Initiative* (*FCI*) network

Besides putting 'super-ageing' on the agenda, FCI has several differences from the Eco-model Cities. First, several experts have 'unofficially' participated in 'visioning' meetings from fall 2010 to spring 2011, searching for a long-term vision for creating world-leading human-centred cities. Second, the government had allocated 1 billion yen yearly for the Initiative for the first 2 years. Third, the project had an international focus to exchange theory and practice for sustainable urban development through participating in international conferences overseas, organizing a yearly International Forum on the FCI in Japan, and supporting each selected FCI for international cooperation. Fourth, CASBEE City[3] was developed, a framework to monitor the quality and performance of the selected "Future Cities" from a triple bottom line perspective (environment, society, economy).

Eco-model Cities was introduced by the Liberal Democratic Party (LDP), which has been in power for most of the post-World-War II period, and the FCI by the Democratic Party of Japan, governing from September 2010 to November 2012. After the LDP returned as governing party in 2012, both initiatives were merged and integrated. At the practice level, the two initiatives were no longer clearly separable after 2014 as they are both run by the Cabinet Office with no separate budgets. Also, the Promotion Council for Low-carbon Cities was merged with the Promotion Council for the "Future City" Initiative.

[3] CASBEE stands for Comprehensive Assessment System for Built Environment Efficiency developed by the Japan Sustainable Building Consortium. CASBEE City is a system that comprehensively evaluates the environmental performance of a city. It also looks multilaterally at the quality and performance of a city from a triple bottom line perspective of the environment, society, and the economy (Japan Sustainable Building Consortium 2012).

3.3.3 Green and Local Autonomy Model City

When the Democratic Party of Japan won the elections in 2010, empowering local municipalities and promoting renewable energy locally were part of their manifesto. This focus makes sense against the background of a highly centralised electricity market and also political structure with its roots in the Meiji Restoration in 1868. The assumption of the government was that reforming these centralised systems would change the stagnating economy and initiate local growing paths (Ministry of Internal Affairs and Communications 2010–2011).

The *Green and Local Autonomy Model City* initiative was set up to collect good practices from frontrunning municipalities that had already pursued local self-reliance for energy, food, and employment. On the basis of these good practises, policies for decentralisation could be formulated in a next step. Higashiohmi (Mizuguchi et al. 2016, Chap. 5, this volume) was selected as one of three municipalities[4] in an informal selection procedure. The Minister of General Affairs, also responsible for promoting local government, asked journalists and consultants to propose prospective cities and, based on this, nominated three (Fig. 3.4).[5]

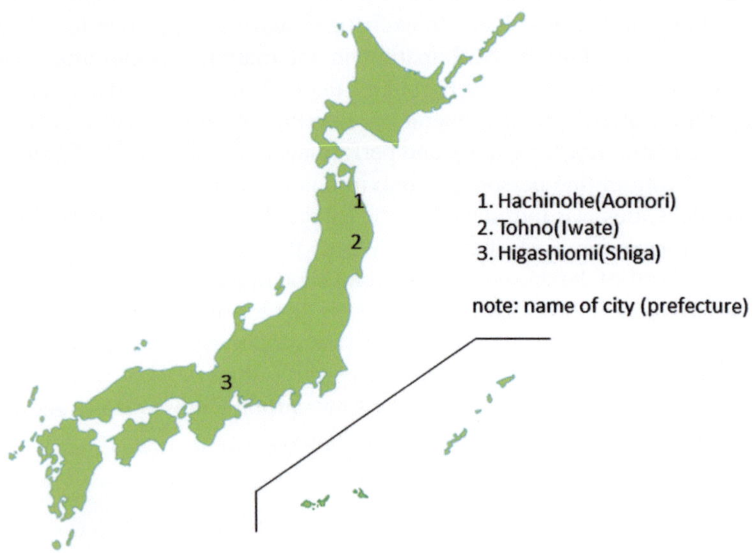

Fig. 3.4 Location map of Green and Local Autonomy Model Cities

[4] The other two cities were located in the northeast region of the main island; both cities were famous for urban and surrounding rural developments based on locally produced food and energy.

[5] Interview with Michiko Yamaguchi, municipality official of Higashiomi, 17 June, 2014, Higashiomi.

Each Model City was allocated 30 million yen for researching and organizing their experience for the committee in 2011 (Ministry of Internal Affairs and Communications 2010–2011). The mayor and middle ranking staff of the three participating cities were invited to join a governmental advisory committee for studying local cases. The local good practiceswere collected, edited, and published by the Ministry and shared with municipalities all over Japan. The program was stopped when the LDP returned to the government. However, it gave the participating cities the opportunity to assess local resources for sustainable development.

3.4 Overview of Cities

All four networks aim(ed) at addressing persistent societal problems such as climate change, an ageing population, and stagnating economies. The main advantages for the participating cities seemed to be learning from each other and the access to funding or to governmental networks (the latter mainly in Japan). Financing these networks allowed the Japanese Government and the European Union to put certain societal problems on the agenda: especially, the Japanese situation shows the vulnerability of these kinds of schemes in terms of the ruling political agenda. MUSIC (at least from the outset), Eco-model Cities and Green and Local Autonomy Model Cities were focusing on specific issues such as a low-carbon society or local resilience, FCI was the most comprehensive and financially well funded. Also, MUSIC was explicitly based on transition thinking and included the implementation of a prescriptive transition management process, none of the three Japanese programs was based on transition management or transition thinking from the outset. However, transition thinking had begun to influence some of the selected cities and Cabinet Office because it was introduced as part of the International Conferences of FCI.

For this book, we selected five cities which were part of either of these networks (Table 3.2). These cities were selected to provide insights into how they addressed different sustainability challenges conceptually and practically by searching for new and innovative methods and instruments based on shared principles of a transitions approach. This selection includes three European cities and their quest for CO_2 emission reductions, namely, Aberdeen, Ghent, and Montreuil of the MUSIC network. These three cities show variety in how a transition management approach can be operationally applied in cities: in terms of timing of the process, openness to transition thinking and approach, as well as framing of the core challenge. For example, in Montreuil, a transition management process, while focusing on CO_2 emissions in the beginning, turned out to be much more on how a city could be governed in a different way. The selection also includes two Japanese cities, namely, Higashiomi and Kitakyushu. Looking at these cities from a transition management perspective, we can identify interesting transition governance activities. In addition, these cities show different transition paths and core transition challenges.

Table 3.2 Overview of the five cities described in this book

	# of inh.	Area size and geography	Core issues	Policy culture	(Transition) challenges for urban development	Focus of (transition) governance activities
Aberdeen, Scotland	210,400	184.46 km² coastal	• Basis for offshore oil industry in the North Sea	• Distrust towards city council	• Ambitious CO_2 reduction goals • Create a culture of energy efficiency among public and private actors	Reframing the challenge, problem understanding and visioning
Higashiomi, Japan (Green and Local Autonomy Model City)	120,000	388.58 km² Center of Japan, adjacent to Japan's largest lake and mountains	• Local self-reliance on care, food, energy and job	• Seven municipalities with their own policy cultures merged in 2008 due to financial fragility • Councils for urban/rural development were formed • Tradition of "do it by local business" culture	• Ambitious CO_2 reduction goals • Job creation on food, energy, care • Super ageing society	Interactions between niches and regimes across different sectors
Ghent, Belgium	240,000	156.18 km²	• Home to third biggest port in Belgium • Many cultural heritage protected buildings	• Municipality and politicians dedicated to climate agenda	• Ambition to become climate neutral • Start of 'climate alliance', a network of diverse committed actors	Broadening and taking action
Kitakyushu, Japan (*Eco-Model Cities* and *Future City Initiative*)	970,000	488.78 km² The northernmost tip of the island of Kyushu, the westernmost island of the Japanese major islands	• The world capital of sustainable development • Low-carbon • Pollution control • Energy efficiency • Internationalization	• Strong leadership of mayor • Strong influence of steel, chemical and automobile industries • Cooperation among citizen, government, industry	• Ambitious CO_2 reduction goals • Job creation on environmental fields • Super ageing society	Toward becoming an international and technological city with waterfront, green environment and human contact
Montreuil, France	103,000	8.92 km²	• Former agricultural and industrial city • Third most populous suburb of Paris	• Great diversity of citizen initiatives promoting environmental, social and economical alternative practices but connection to city council missing	• Ambitious CO_2 reduction goals • Create a culture of efficiency for a 'positive energy strategy'	Transition within policy makers and civil servants' cultures and practices

References

Cabinet Office (2008) Application handbook of eco-model cities (proposal). Cabinet Office, Tokyo (only in Japanese)

Cabinet Office (2013) "FutureCity" initiative, mobilizing future urban planning. Cabinet Office, Tokyo (only in Japanese)

Frantzeskaki N, Tefrati N (2016) A transformative vision unlocks the innovative potential of Aberdeen City, UK. In: Loorbach D, Wittmayer J, Shiroyama H, Fujino J, Mizuguchi S (eds) Governance of urban sustainability transitions, pp 53–72. Springer, Tokyo

Hölscher K, Roorda C, Nevens F (2016) Ghent: fostering a climate for transition. In: Loorbach D, Wittmayer J, Shiroyama H, Fujino J, Mizuguchi S (eds) Governance of urban sustainability transitions, pp 95–115. Springer, Tokyo

ICLEI (2012) Local sustainability 2012: taking stock and moving forward. Global review. ICLEI Global Report, Freiburg

Interreg IVB NWE (2013a) http://www.nweurope.eu. Accessed 20 Dec 2013

Interreg IVB NWE (2013b) NWE in action! INTERREG IVB North-West Europe. Programme overview 2007–2013. 12 reasons to engage. INTERREG IVB NWE Joint Technical Secretariat, Lille

Japan Sustainable Building Consortium (2012) CASBEE for cities. Technical manual, Tokyo

Krauz A (2016) Transition management in Montreuil: towards perspectives of hybridization between 'top-down' and 'bottom-up' transitions. In: Loorbach D, Wittmayer J, Shiroyama H, Fujino J, Mizuguchi S (eds) Governance of urban sustainability transitions, pp 137–154. Springer, Tokyo

Ministry of Internal Affairs and Communications (28 April 2010 to 1 March 2011), meeting materials for Green and Local Autonomy, Ministry of Internal Affairs, Tokyo (only Japanese are available). http://www.soumu.go.jp/main_sosiki/kenkyu/bunken_kaikaku/index.html. Accessed 15 May 2015

Mizuguchi S, Ohta K, Beers PJ, Yamaguchi M, Nishimura T (2016) Interactions among multiple niche-innovations and multi-regimes: the case of the "Welfare Mall" in Higashi-Ohmi. In: Loorbach D, Wittmayer J, Shiroyama H, Fujino J, Mizuguchi S (eds) Governance of urban sustainability transitions, pp 73–93. Springer, Tokyo

Murakami S (2012) Special report on environment and energy, 19 June, 2012 issue of Nikkei (Japan Economic Daily), Japan Economic Daily, Tokyo

Promotion Council for the "FutureCity" Initiatives (2014) "FutureCity" initiative. Cabinet Office, Tokyo

Roorda C (2013) Learning in a transnational project: the case of MUSIC. In: Accelerating change – delivering sustainable energy solutions; proceedings of the joint seminar by Intelligent Energy Europe and Interact: 11. http://ec.europa.eu/energy/intelligent/files/library/brochures/accelerating-change-delivering-sustainable-energy-solutions.pdf. Accessed 6 June 2015

Shiroyama H, Kajiki S (2016) Case study of Eco-town Project in Kitakyushu: tension among incumbents and the transition from industrial city to green city. In: Loorbach D, Wittmayer J, Shiroyama H, Fujino J, Mizuguchi S (eds) Governance of urban sustainability transitions, pp 117–136. Springer, Tokyo

United Nations Institute for Training and Research (2015) Towards a post 2020 agreement. United Nations Institute for Training and Research, Geneva

Wittmayer JM, Loorbach D (2016) Governing transitions in cities: Fostering alternative ideas, practices and actors. In: Loorbach D, Wittmayer J, Shiroyama H, Fujino J, Mizuguchi S (eds) Governance of urban sustainability transitions, pp 17–36. Springer, Tokyo

Zelinsky W (1991) The twinning of the world: sister cities in geographic and historical perspective. Ann Assoc Am Geogr 81(1):1–31

Part II
Transition Management in European and Japanese Cities

Part II is the heart of the book and includes five rich empirical chapters describing cities and their quest for sustainable development. What holds these chapters together is their engagement with the transition management approach. The chapters on the Japanese cities (Chaps. 5 and 7) describe urban governance analysed through a transition management lens and the chapters on the European cities (Chaps. 4, 6, and 8) outline the processes and outcomes of an operational application of transition management, where co-creative processes between city administrations and other urban actors were organized based on a transition management methodology.

All chapters are structured along a basic rationale, introducing first basic background information on the city and its main sustainability challenges and the main analytical framework and/or focus. This is followed by a description of the governance mechanisms addressing the challenge and their outcomes. The chapters close with an analytical discussion and conclusion.

Chapter 4
A Transformative Vision Unlocks the Innovative Potential of Aberdeen City, UK

Niki Frantzeskaki and Nora Tefrati

Abstract The breakthroughs in terms of thinking, framing, and unlocking Aberdeen's potential were realised during the envisioning phase of a transition management process. As thus, we zoom in on the envisioning phase and elaborate its outcomes as being the 'critical phase' in which diverse interests, perspectives, hidden assumptions, and ingrained perspectives were expressed, negotiated, and debated. This action resulted in a new understanding of the city's persistent problems and a holistic perspective for future sustainability in the city. From our analysis, the successful output of the envisioning process was an integrative understanding of future aspirations that comply with sustainability values (intergenerational justice, fairness, equity) and, in turn, created a momentum for inclusive social dialogue and mobilised networks to take the vision into practice.

Keywords Transition management • Envisioning • Energy security • Inclusivity • Vision • Sustainability

4.1 Introduction

Aberdeen is the oil-energy capital of Scotland. From the historical transition, from a fishermen's village to an oil capital, not all citizens in Aberdeen enjoyed the benefits of the industrialisation. It now appears that the social and financial vulnerabilities of a new transition away from an oil-dominated economy can follow the same pattern with the risk of the benefits being unequally distributed (Aberdeen City Council 2009, 2010, 2011).

Given this challenging present, Aberdeen has experimented with transition management as a governance approach to aid the search of alternative pathways towards urban sustainability and to investigate how to mobilise collaborative arrangements between different actors in the city (Roorda et al. 2014). During the

N. Frantzeskaki (✉) • N. Tefrati
Dutch Research Institute for Transitions, Erasmus University, Rotterdam, The Netherlands
e-mail: frantzeskaki@drift.eur.nl

D. Loorbach et al. (eds.), *Governance of Urban Sustainability Transitions*, Theory and Practice of Urban Sustainability Transitions,
DOI 10.1007/978-4-431-55426-4_4

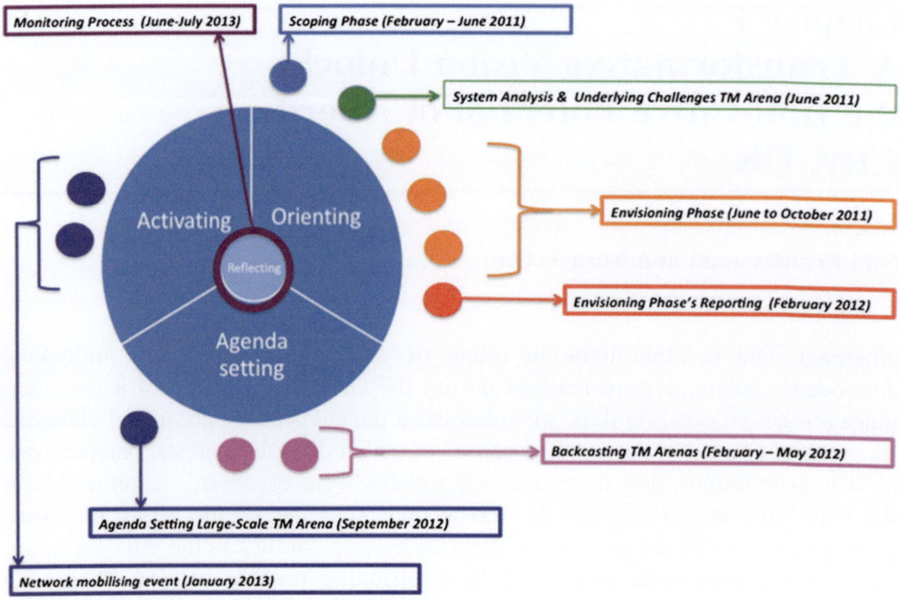

Fig. 4.1 The transition management process of Aberdeen city as realised with the MUSIC project (*Source*: Authors)

scoping phase of the transition management process, context dynamics were investigated and actor mappings were realised in preparation of the process (Fig. 4.1). The authors engaged with policy, corporate, and civil society change agents in Aberdeen, discovering that there are overarching local needs of the city in their strive for urban sustainability and resilience. The three most prominent needs are the following.

- There is no legitimised and socially embraced long-term sustainability vision in the city. The recent sustainability vision received the acceptance of the policy officers internally; it nonetheless failed to inspire and guide action at larger city scale because of the way the process for formulating the sustainability targets was realised.
- Fear of loss or declining energy security emerges as a marginal narrative within the local administration. As such, it creates leverage for considering aspects of social justice and equity issues, with one of the policy officers stating "leave no one behind in this transition" or "move away from having a tale of two cities" (Aberdeen Policy Officer Interview 2011).
- There is a voiced need for inclusive long-term action towards energy security and energy resilience in the future that considers new business context (small–medium enterprises, local businesses, renewable energy corporations) while retrofitting contemporary energy businesses to a new economy. As a policy officer stated: "How to utilise the expertise of people who now work at the oil industry locally at the post-oil era?" (Aberdeen Policy Officer Interview 2011).

Considering such imperatives for action, a transition management process funded and facilitated by the MUSIC project research team (see Chap. 3) opened up a dialogue between policy officers, change agents from the city, scientists, and community representatives to rethink current and future pathways for Aberdeen. The research question that our contribution addresses is this: *Can participatoryenvisioningenable the co-creation of asustainabilityoriented vision to inform strategic environmental planning while steering clear from historically and culturally entrenched practices and perceptions?*

In this chapter we focus on the process of participatory vision building and its outcomes in Aberdeen that is the second process phase of the transition management cycle (Loorbach 2010; Frantzeskaki et al. 2012a; Nevens et al. 2013). The MUSIC project has had the overall objective to work on a long-term strategy by applying transition management (TM) as the (social, policy, business) change agents' consultation methodology for developing scenarios and strategic pathways to gain insights into opportunities for transforming energy urban planning and governance.

Having completed the TM process in Aberdeen (lasting from 2011 until 2013), and with the city obtaining the Scottish Green Apple award in 2014 for its innovative governance approach of adopting TM, we choose to elaborate on the envisioning phase because during this phase the unlocking of entrenched perceptions and assumptions about current and future sustainability of the city of Aberdeen occurred. It is during the envisioning phase that the change agents via facilitated dialogue and open deliberation of concerns, criticisms, and ideas unlocked from historically bound perceptions of Aberdeen as an energy city and recognised that daring action is needed to achieve inclusive future sustainability. The unlocking and realisations that occurred during the envisioning phase and the co-creation of the transformative vision for Aberdeen paved the ground for thinking about strategic actions that holistically orient and change the city towards sustainability.

We present a visual overview of the TM process in Sect. 4.2. For every step of the envisioning phase, we outline the methods we employed in Fig. 4.2. In the following sections, we present the conceptual model for designing and evaluating transition-oriented envisioning (Sect. 4.3), the outcomes of the envisioning process applying the conceptual model, and the lessons learnt from monitoring and evaluating the envisioning process (Sect. 4.4). We conclude the chapter by addressing the research question via examining the role and contribution of the envisioning process in the context of Aberdeen City (Sect. 4.5).

4.2 Overview of Aberdeen's Transition Management Process

The TM process in Aberdeen City included in total ten transition arena meetings, participatory workshops following the transition management process methodology, as mapped in Fig. 4.1.

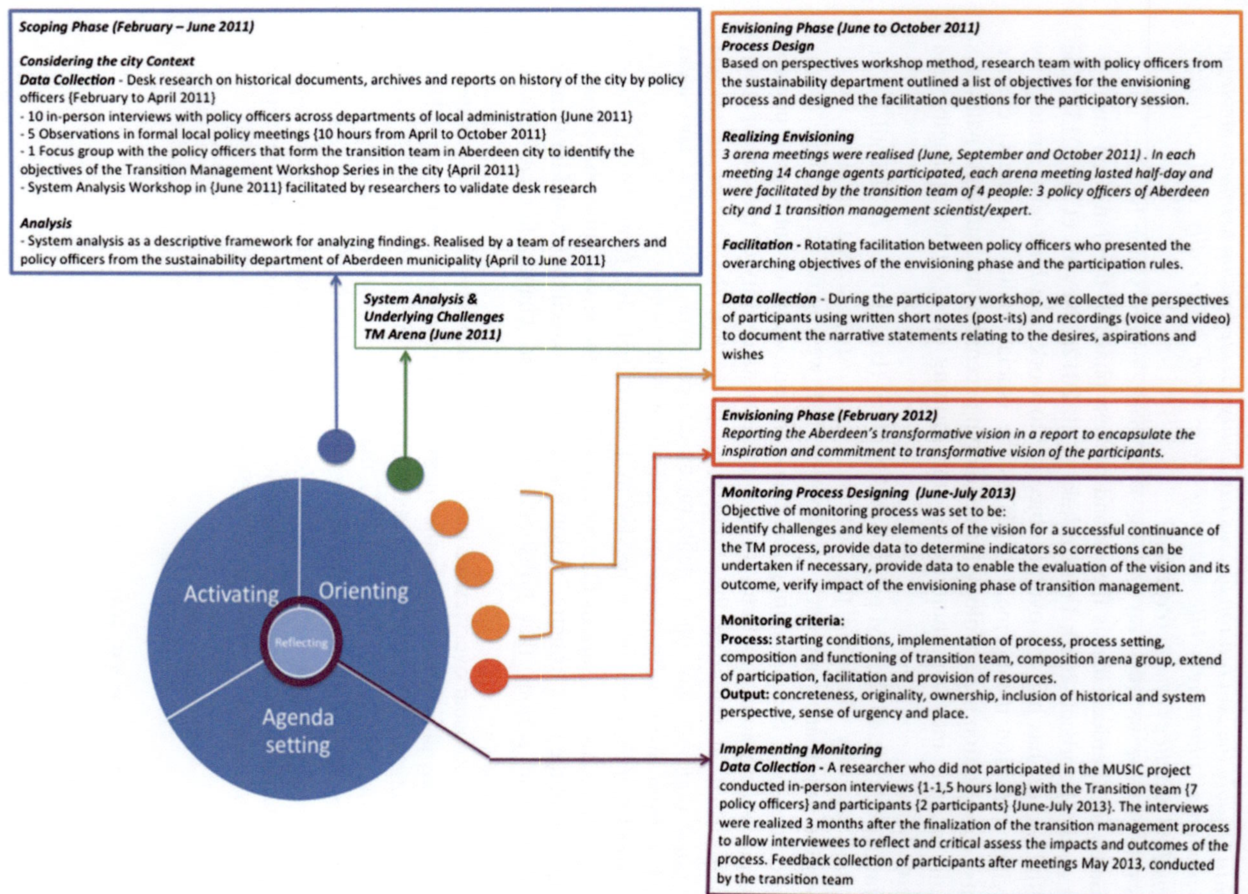

Fig. 4.2 Detailed presentation of the envisioning phase and information on the methods used in preparing each transition arena meeting, collecting data, and for facilitation of each transition arena meeting in the city of Aberdeen (*Source*: Authors)

4.3 Envisioning: Co-creating Transformative Visions of Sustainability

As part of the TM process, the envisioning step is placed as one of the four process handles of the Loorbach's (2010) TM cycle.

4.3.1 Designing Transition Management Processes

For designing TM processes, three components of current and future dynamics need to be taken into account: context, process, and outcomes. These three cornerstones constitute a new conceptual lens on the Transition Management Process Design Framework (Fig. 4.3) (Frantzeskaki et al. 2012a, b). The TM design framework considers three interconnected aspects.

(a) Process: The operational methods (e.g., dialogues, debate, creative scenario method) needed to fulfil objectives and to deliver specific expected outcomes that relate to content.
(b) Content: The desired outcome from the process and the form of this outcome (e.g., report, presentation, visualisation, and shared storyline, strategy assessment). The content pillar includes the effort of customizing the outcomes to the context, taking into account policy and social needs as well as timing.
(c) Context: The process needs to be tailored to consider, align, and tap into context dynamics. As thus, context is considered in the overall design by negotiating and co-creating content to context as well as by continuous

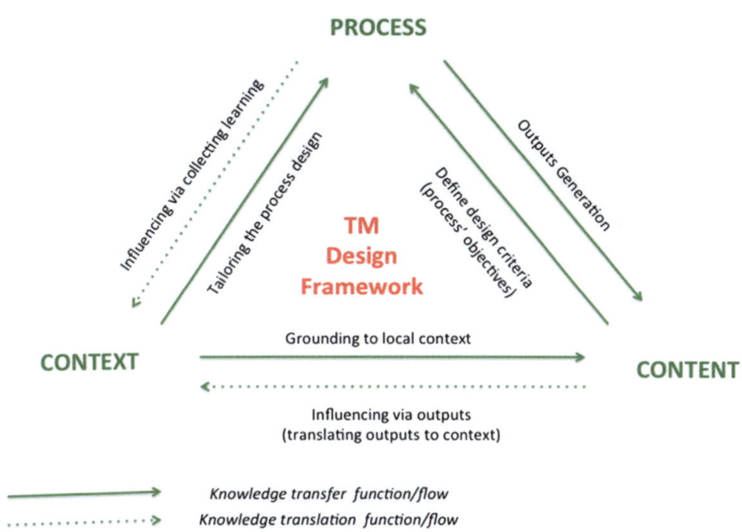

Fig. 4.3 Transition management process design triangle (*Source*: Authors, Frantzeskaki et al. 2014b)

monitoring and assessment of context dynamics to the process. In this way, context dynamics inform process decisions (e.g., methods of engagement) and content and, at the same time, content and process can transcend context dynamics and (may) create leverage points for transformative interventions.

4.3.2 *Envisioning Phase of Transition Management Process*

Envisioning is a process of imaginary scenario building. Envisioning is a process that takes place in an arena setting and involves actors from local communities. Envisioning is therefore a process of engaging community and local change agents. The vision is the starting point for initiating sustainability transitions (Newman and Jennings 2008). In this process of voicing and acknowledging desires, dreams, and wishes to construct a common future image that will link and inform search for action, active participation of actors with a stake and 'future take' on the issue or system is essential. Local stakeholders are important for the envisioning process, and their active participation and engagement can increase both the accountability of the process outcomes as well as the successful implementation of the derived plans (Carlsson-Kanyama et al. 2008; Vural-Arslan and Cahantimur 2011). The way local change agents and stakeholders are involved is crucial for the accountability and the legitimacy of the envisioning process (Larsen and Gunnarsson-Ostling 2009; Larsen et al. 2011). One of the direct benefits of participatory envisioning is the direct dialogue between local stakeholders and policy makers as well as the direct access to policy that local stakeholders are given (Larsen et al. 2011, p. 417).

Some content characteristics and objectives of a vision include the following:

- Visions provide a firm foundation for policy formulation because in an envisioning process "awareness is raised, the public is motivated to become involved, a sense of identity is fostered within the city, everyone's view is valued, partnerships are formed that can assist implementation, and a basis for conflict resolution is established." (The Cities Alliance 2007, pp. 24–25)
- The vision depicts the desirable characteristics of the area distinctive of its nature and characteristics and considering the 'sense of place' (Newman and Jennings 2008, p. 8).
- The vision includes the shared aspirations of the participants addressing sustainability (Newman and Jennings 2008).
- Understanding and fore-facing the desirable directions or images that are related to sustainability (of the city or region) (Beers et al. 2010).
- A vision brings forward the potential of the city or region as represented/mediated by local stakeholders (van Eijndhoven et al. 2013).

Some process and learning objectives of envisioning include the following:

- To communicate climate pressures to local contexts or, as Sheppard et al. 2011 (p. 410) put it: "Climate change scenarios can be spatialised at the local level to allow analysis of climate change impacts, vulnerabilities, and adaptation/mitigation suitability, making them more likely to be integrated into planning processes."

- To sustain an open dialogue environment that allows different knowledge types to come together: Scientific knowledge in terms of future projections (e.g., climate change scenarios), local knowledge on contemporary processes, and tacit knowledge of practitioners into the change potential and resisting elements in a societal (sub)system (Sheppard et al. 2011).
- To encourage and promote new ways of thinking about the future and to create collaborative ties between previously unconnected or disconnected actors (Berkhout and Hertin 2002; Ozkaynak and Rodriguez-Labajos 2010).
- Engage people authentically and provide space for reflection and creativity (Newman and Jennings 2008, pp. 12–13) that, at the same time, paves the ground for change given that facilitated reflection and enacting creativity challenge "conventional wisdom and encourage debate" (Ozkaynak and Rodriguez-Labajos 2010, p. 996).

An envisioning process includes three steps:

(a) Formulating guiding principles
(b) Creating the vision
(c) Operationalizing vision elements into strategic objectives

For each of these steps, we elaborate on the three design elements (process, content, context) based on scholarly work from foresight science, scenario approach, and operations research (Bishop et al. 2007; O'Brien and Meadows 2013; Ringland 2010; Robinson et al. 2011).

(a) Formulating Guiding Principles
Process: The objective of this step is to formulate a suite of principles that represent desired system outcomes for the long-term horizon. In this step of the process, core values of the participants surface and are negotiated to formulate principles that work in synergy and are to guide future developments (Schultz et al. 2007; Rogers and Bazerman 2008).

Content: The expected outcome of this step is a list of guiding principles that is the base of the broader vision and which will breed the vision-building process.

Tailoring content to the context: The way that the guiding principles are to be presented to the participants can differ based on whether the information sharing between the participants and non-participants is open or closed. An overview of the negotiated guiding principles along with a short description of their meaning can suffice to bring the envisioning process forward. However a careful presentation is needed if the guiding principles are to be communicated or shared with non-participants or in forums and processes outside the current process (Pichert and Katsikopoulos 2008).

Context: To consider context dynamics, it is suggested to (a) search, acknowledge, and present existing visioning work, especially when there are participants in the envisioning process who are aware of or participated in currently terminated or ongoing parallel envisioning processes; (b) take into consideration recent developments including pilot projects, emerging initiatives from communities or partnerships, as well as existing (sustainability) agendas or planning programs; and (c) provide the choice to the participants to either build on existing work or start with a new perspective to build principles.

(b) Creating the Vision

Process: The objective of this step is to create storylines or images of the envisioned future system to express agreed future desires and wishes upon a system or situation. The successful vision captures the imagination of both participants and of a broader audience and can create symbolic value in a system or organisation (Shipley 2000; Shipley and Newkirk 1999; O'Brien and Meadows 2013).

Tailoring the process to context dynamics, scholarly work on envisioning and creative scenario building, offers a great variety of empirical examples but limited knowledge on operational guides for tailoring. Among the limited and fragmented 'good practices' on contextualisation of envisioning processes are (a) to consider existing narratives around change and trigger them by designed 'crises' contexts within the vision (Sondeijker et al. 2006; Wiek et al. 2006); (b) to allow for open confrontation and open search of commonly shared values and future desires by engaging with a variety of actors, or simply allow for not-like-minded people to envision together for a vision (Helling 1998; Van der Helm 2009); and (c) to include different knowledge representatives into the envisioning for co-creation and learning (Sheppard et al. 2011; Gidley et al. 2009).

Content: The expected outcome from this process step is a comprehensive description of a vision that outlines and synthesises different images or representations of the desired future. The overall vision can be presented in having different themes or different images. The vision can be presented in the form of storylines, artistic impressions and/or expressions (e.g., images or videos or sculptures), headlines, or front pages of future newspaper or newsmagazine issues.

For a vision to create social value or simply, symbolic value in its context, it has to transcend from the group that created it and become relevant and dialectically absorbable by the context (Hughes 2013). Communication of the vision from the group or group representatives to a broader audience or targeted audience is a side step that may enable value creation and mobilisation of networks and resources for realisation of the vision (Hughes 2013; Tompkins et al. 2008; Frantzeskaki et al. 2014a; Volkery and Ribeiro 2009).

(c) Operationalisation of the Vision into Strategic Objectives

Process: The objective of this step is to generate indicators (strategic objectives) that relate to and are operational translations of the guiding principles. The role of the strategic objectives is to create a suite for assessment of action: how to assess if action taken brings the system closer to the vision, or simply: how to know whether a vision is achieved.

With this comes a reflection that achieving a vision is not a goal in itself. At the same time, having a suite of strategic objectives provides the means to better design short-term actions to open the route for more daring and transformative steps on the medium and long-term.

Content: A suite of strategic objectives is to be formulated focused on the values that are represented in the guiding principles and the vision description (Keeney 1996a, b; Frantzeskaki and Walker 2013).

Context: Taking context dynamics into account for operationalising the vision implies that existing indicator schemes may be incorporated or partially adapted if these fit the overall vision content. At the same time, context dynamics may drive the operationalisation 'dialogues' and 'search processes' into a feasibility-trap and tone down the radicality of the vision. At this step, context dynamics need to be carefully considered so as to have a constructive meaning and influence rather than a disorienting or distracting impact.

4.4 A Transformative Vision for Aberdeen's Sustainability

4.4.1 Formulating Guiding Principles

Process Design The Aberdeen MUSIC project team realised a perspectives workshop where the transition arena participants first reflected on the formulation of the guiding principles and then asked how to realise the guiding principles. The output of the perspectives workshop is structured by the project team and is presented in the following paragraphs.

The first envisioning arena meeting was held in June 2011 and the second in September 2011. During the first and second arena, participants were asked to raise and identify the principles that will guide transformative action. The guiding principles express the desires of the group, enable a focus or common direction and work in conjunction with one another with the aim of achieving a sustainable Aberdeen. The participants were asked to express and explain their responses to the following questions:

What would Aberdeen be if it were your dream sustainable city?
What would you want Aberdeen to be in the future?

A third envisioning workshop was held in May 2012 to introduce new frontrunners to the process and to gather their thoughts on the guiding principles to date. The main outcome of this workshop was that the guiding principles needed to be more personalised with a more visible green thread.

The MUSIC transition team consolidated the responses of the transition arena participants into a list of guiding principles for a sustainable Aberdeen. These guiding principles were revised and enriched during the third workshop (held in May 2012). The group identified two time horizons that relate to the vision and possible transition pathways: 2030, to have an urgency of doing things that can be done (medium term), and 2050, as a time-mark to be prepared for when the oil industry has left Aberdeen (long term).

Outcomes The group formulated five guiding principles and created five vision images for Aberdeen's sustainable future. We present below the synthesis of the envisioning arenas after consolidating the outputs of the three envisioning workshops (Table 4.1).

Table 4.1 Guiding principles of a sustainable transition in Aberdeen city

1. Aberdeen as an opportunity city
In 2050 we will continue to attract and generate investment, business, and offer a diversity of employment prospects without compromising emission reduction targets
2. Aberdeen as an attractive city to visit and live
In 2050 we have a vibrant city boasting a multitude of cultural places, which are in balance with nature, citizens enjoy living here, and take pride in their city
3. Aberdeen as a leaning city
In 2050 our city has strong international connections and is a centre of excellence and an attractor for knowledge and expertise on energy, medicine, ecology, and business where people come to learn from its examples and achievements. Lifelong learning and reliable ICT infrastructure and technology ensured sustainable work practice
4. Aberdeen as an accessible city
In 2050 all neighbourhoods in our city are accessible to all via an integrated walking, cycling, and public transport network, with the city centre designed for all mobility needs
5. Aberdeen as an energy-efficient and resilient city
In 2050, our city is an energy capital built on its oil and gas era that is resource efficient and has energy supplied from sustainable energy sources. We are the leading low carbon city in Scotland

Context By using several expressions for each principle (e.g., we continue, we are, we have), it is apparent that the participants have seen the creation of guiding principles as an opportunity to connect with ongoing activities and existing projects as well as to build up new project ideas for achieving sustainability. By committing to the guiding principles, all participants could link their interests, desires, and expertise to at least one of the principles.

By referring to ongoing activities of the city (e.g., sustainability plans, ongoing initiatives), change agents enriched the guiding principles and acknowledged topics and actions that have an indirect impact on carbon curbing despite the concern of the policy officers involved that the focus on CO_2 reduction had been lost. The participants – local change agents – not only could demonstrate a good understanding of sustainability but also fore-facing the desirable directions (or vision images) that relate to sustainability of the city and the region (Table 4.2).

4.4.2 Vision Creation

Process Design After agreeing upon the principles that are set to guide the sustainability transition of Aberdeen City, the transition team and the change agents worked in smaller groups on different vision images for each guiding principle in a written form. The change agents chose the guiding principles with which they could identify the most. While they were working in small groups and discussing the outline of the vision image, an artist attended the meeting to gather input for the creation of appealing images. His task was to capture the discussions of the

Table 4.2 Relating the guiding principles to the context

Coherence	The guiding principles are connected to the themes discussed at the problem definition
Concreteness	The guiding principles describe a well-rounded picture of Aberdeen in 2050 with a reference to a never-ending transition but remain on quite an abstract level
Inclusion of system perspective	The guiding principles include the traits of sustainability in terms of ecological, economic, and social aspects
Originality	The different guiding principles are original in terms of identifying issues which are not directly connected to the main topics of oil dependency and CO_2 neutrality. Although the need for a fundamental chance has been discerned, the principles do not much vary from the status quo and aim more to enhance technical innovations
Relation to target population	One original issue was to bundle Aberdeen and Aberdeenshire to imply more people to the sustainable transition and to grasp the challenges of the city from a holistic point of view because the carbon issue does not know any geographic boundaries
Ownership	By having persuaded the change agents to achieve a consensus about the guiding principles as the precondition for further collaboration, the ownership could be attained and the local actors are authentically engaged to the process

participants in images and sketches for both involved actors and the broader audience.

Outcomes The expected outcome of this step is to create a comprehensive description of a vision and to outline the desired future by highlighting the potential of the city and its distinctive nature. We present below the respective output of the vision creation by listing the vision image of each guiding principle (Table 4.3).

Sketches and images for each guiding principle, for the transition process as understood and discussed by arena participants and for the future actions in general, were designed by the artist. The sketches and images were used along with the process to illustrate internally and externally the core ideas of each vision image. We present here the illustration most published by the city council in electronic bulletins, reports, and during city conferences. Figure 4.4 is not only representing the content of the multi-actor dialogues during the transition arena workshops but reflects the atmosphere of the discussions as complicated and direction changing upon searching and seeking the desired future of Aberdeen City.

Context Because the images were produced during the meetings and took up the conversations, they included the different perspectives of the participants. No single scenario for Aberdeen in 2050 was created during the meetings, because like-minded people who opted for diversity of ideas rather than polarisation of opinions did not attend the meetings. This fact proves that the selection of change agents during the scoping phase fulfilled the criteria of having a successful co-creative process to discover transformative potential for sustainability. The vision images

Table 4.3 Transition vision and vision images

Vision image 1	Aberdeen continues to attract and generate investment and builds on the achievements being fulfilled through having an active business sector. The transition to an opportunity city in 2050 is an evolution of the current situation. In 2050 Aberdeen has a strong economy not reliant on the oil industry. The city maintains its position as the centre for worldwide oil and gas subsea expertise while widening its R&D expertise and training in universities. The skills and knowledge developed during the oil and gas era are retained and applied to renewable technologies. At the same time, investments and diversity of opportunities and employment prospects allows for an increased awareness of a more diverse economy. Strong ties exist between industry, research, and local authority that has created a change of mindset about the singular nature of employment in Aberdeen. This change of mindset was the outcome of coordinated actions and information made available to the communities that resulted in making them aware of the city situation about employment and where this employment originates. In 2050 citizens are aware of city opportunities and actively involved in the reality of the city. The window of opportunity for the post-oil change has been seized, and the city is more European oriented in seeking opportunities for cooperation and learning
Vision image 2	In 2050, Aberdeen City provides a high quality of life for all that live in the City. Aberdeen's city centre is a vibrant and attractive place. Aberdeen's City Centre Development Framework quarters have been realised with the city centre designed for people. Flagship projects and ventures will bring more visitors to the city, alongside this Aberdeen's granite heritage and distinctive aesthetic is celebrated. Heritage, arts, culture and leisure sites within the city centre are easily accessible and connected to an integrated public transport system. Aberdeen has affordable and well-designed housing, in collections of urban villages where communities flourish, services are provided, and there is local economic growth and attractive neighbourhoods. Aberdeen is a tourist destination and offers a wide range of heritage, sport, and leisure activities. Aberdeen City is the tourist gateway for visitors to access the Cairngorms National Park and the River Dee
Vision image 3	In 2050 Aberdeen City is a learning city and new disciplines are active and are developed such as public health, energy, environmental sustainability and technology studies. The knowledge institutions are connected to the industry and to the development of the city's employment reality. There is a strong collaborative work ethic with other cities and institutions throughout Europe. In 2050 the schools in Aberdeen offer a high quality curriculum and families are attracted to come to Aberdeen for the learning/education opportunities found in place. International schools are also operating and attracting expatriate communities. At the same time, universities and colleges are performing high-quality research that is communicated globally. Aberdeen is a destination as a learning attractor for postgraduate research and a learning city with long-life learning programs. There are strong links between the public–private institutions in education and employment. As a modern learning city, modern technology is at a place that helps both long-life learning, as well as offering education beyond conventional channels and domains and accessibility to knowledge for all. In this learning environment, citizens work more efficiently and also from home
Vision image 4	In 2050 Aberdeen is a highly interconnected city while having pedestrian areas that all citizens enjoy. The mobility network of the city is well organised, with well-planned connected hubs. Walking, cycling, and public transport modes are prominent with improved routes and lanes in neighbourhoods and throughout the city, and streets have been reclaimed from the car. The city centre is a hub of the

(continued)

Table 4.3 (continued)

	mobility network and is the heart of the city. There is strong use of public transport with schedules that mirrors the work pattern and reflects demands of the citizens. Private, single-occupancy car travel is discouraged with disincentives by the introduction of emissions restrictions. The mobility sector is realised because of the awareness that it is demand driven while also embracing new technologies and new practices. In 2050 Aberdeen City is a carbon-neutral city when considering the public mobility options
Vision image 5	In 2050, Aberdeen City is a leading example for energy efficiency and sustainable energy. There is a well-established forum for best practices and more joined-up networks providing an overview of what is happening in the city. All new buildings are carbon neutral with existing and historic buildings highly insulated and energy efficient. The City has a decarbonised and decentralised electricity supply with 100 % of energy provided by renewable energy from micro-renewables and the City's Offshore Wind Development Centre. The citywide district-heating network provides heating for all private, Council, and social housing as well as public buildings. The city has eradicated fuel poverty. The City is connected by broadband communication networks with flexible and home working reducing the need for travel. The City's integrated transport network promotes active travel (walking and cycling) with the public transport system provided by electric and hydrogen vehicles

created a symbolic value for the group: the group expressed the collective wish to collaborate on other projects in future.

4.4.3 *Operationalisation of the Vision into Strategic Objectives*

Process Design After finalizing the vision images, the vision operational aims were (a) to describe the current situation of the guiding principle of their interest, (b) to identify the strategic objectives in terms of desirable characteristics of the guiding principle, and after that (c) to detect which strategies can achieve the guiding principle in the long term or medium term. The identification of potential first actions is essential for the change agents to identify possible areas for contribution through their own profession at early stages. At a later stage, the identification of first actions will enable change agents to connect with networks that are already working on similar topics as well as to mobilise resources towards achieving the defined sustainable outcomes (guiding principles).

Outcomes The respective outcome of the operationalisation phase are listed next and again structured along the guiding principles (Table 4.4).

Context As already mentioned, the division of the future vision into images that are thematically distinct but work in synergy with each other helps the change agents to categorise a distant future. In this way, it becomes possible for the change agents to come up with feasible work plans and strategies. Subsequently the mode

Fig. 4.4 Vision sketch that interprets the sustainability transition pathways of Aberdeen city

of thinking about the city's sustainable future can therefore be brought from the reflective to the operational level, and involved actors can specifically identify ongoing activities and projects.

4.5 Governance Value for Envisioning plus Lessons for Governance of Urban Sustainability Transitions

The dialogues between planners, policy officers, business and community entrepreneurs, and scientists from local universities allowed us to reestablish relationships of trust, idea sharing, and knowledge co-creation as well as collaborative work towards an inclusive vision. To assess the value of the envisioning process, we analysed the process and content as well as how these two relate to the current

Table 4.4 Strategic objectives for each transition vision image and guiding principle

Guiding principle 1	Aberdeen as an opportunity city
Strategic objectives	• Broaden horizon of opportunities • Open communication channels • Create and support diversity of expertise
Potential first actions	• Assess the expertise we have (oil, food, fishing, learning, intellectual property) and focus on turning into businesses • Build a foundation for green growth • Maximise social business opportunities in energy efficiency • Collaborate in partnership with universities, research institutes to create a knowledge hub • Engagement and partnership working between sectors • Increase in housing to meet citizens needs that is affordable
Guiding principle 2	**Aberdeen as an attractive city to visit and live**
Strategic objectives	• Sustain a diverse cultural economy from grassroots projects • Promote change of cultural attitudes about energy use • Foster a higher sense of pride in the city • Become more creative about hybrid industrial space • Increase variety of cultural spaces and access between them
Potential first actions	• Pedestrianised areas • Behaviour change—respect for the city centre • Look at city centre spaces with potential to transform to attractive spaces • Access funding to provide what is needed for city's promotion
Guiding principle 3	**Aberdeen as a learning city**
Strategic objectives	• Improve research activities from local knowledge institutes • Strengthen knowledge capital of future citizens • Promote diversity of curricula offered by knowledge institutes • Create a city campus network
Potential first actions	• Create a partnership between the renewables, oil and gas sectors, and the research centres • Introduce relevant courses in universities about new energy economy
Guiding principle 4	**Aberdeen as an accessible city**
Strategic objectives	• Improve connectivity between locations and between mobility modalities • Improve bus services and infrastructure • Ensure affordability of mobility options • Expand cleaner mobility infrastructure • Increase awareness about the existing capacity for behaviour change • Improve accessibility for diverse uses
Potential first actions	• Assessment of public transport, walking, and cycling opportunities • Build dedicated cycle lanes • Create end-to-end bus lanes to get people out of their cars • Identify bottlenecks and do something about them
Guiding principle 5	**Aberdeen as an energy-efficient and resilient city**

(continued)

Table 4.4 (continued)

Guiding principle 1	Aberdeen as an opportunity city
Strategic objectives	• Increase zero-carbon buildings • Introduce alternative funding for energy needs • Increase the percentage of RES in the energy mix
Potential first actions	• Build absolute, binding commitments to meet the Climate Change (Scotland) Act 2009 • Establish learning alliances/networks that share experiences and plans • Profile households that have made the low carbon move • Support heat network expansion and conversion to low carbon

context by examining the following points (Loorbach 2010; Frantzeskaki et al. 2014a):

- The strategic value of the vision, in terms of creating legitimacy to long-term goals, values, new culture, and symbols that relate to local identity;
- The tactical value of the vision, in terms of connecting values to mid-term targets, inspiring and mobilizing resources, networks, and partnerships;
- The operational value of the vision, in terms of connecting long-term to short-term actions and projects;
- The reflexive value of the vision, in terms of operating as a means to assess and evaluate existing paradigms, mindsets, wisdom, and routines.

The breakthroughs in terms of thinking, framing, and unlocking the city's potential were realised during the envisioning phase of the transition management process in Aberdeen. From these empirical grounds, we further distilled that an envisioning process has a fourfold governance value.

Strategic Value of Envisioning: Envisioning Dialogues Can Create Space for Detangling Socially Persistent Problems and Uncovering Innovative Potential of Cities An envisioning process when set up and adapted to the context can *create space for detangling socially persistent problems and uncovering innovative potential of cities*, even when sustainability concerns are staggered by social injustice and econocratic myopia. From our experience in Aberdeen's transition management process, the successful output of the envisioning process was an integrative understanding of future aspirations that comply with sustainability values (intergenerational justice, fairness, equity) which created a momentum for inclusive social dialogue and altered ingrained perceptions but also mobilised networks to take over the vision into practice.

At the same time, the main sustainability challenges were identified but not addressed in the vision and the guiding principles. The deliberations of the working groups are dominated by technical solutions, but the process itself enabled them to identify a change of roles of the participants and the council to work towards the desired future in terms of having the city facilitating further activities.

Tactical Value of Envisioning: Vision as an Attractor and Mobilizing Icon for Networks of Actors to Collaboratively Break Through Current Patterns of Practice That Hold Back Innovation The envisioning process yielded a new transformative vision for the context of Aberdeen city with feasible elements. Contrasting the vision to the current situation shows the space for action that exists to achieve the future vision. In this way, the vision *acts as an attractor or mobilizing icon for networks to collaboratively break through current patterns of practice that hold innovation latent* (the attracting power of the vision).

Vision as an Expression of Political Empowerment of the Change Agents' Group The political empowerment relates to the ability of the process development and the actors involved to create new ties among the frontrunners as well as between frontrunners and the transition team. Because the arena process is still in progress and both change agents and transition team confirmed the added value of the process in terms of getting in contact with people who share the same interest and contribute with their own knowledge, they still try to engage new actors, organisations, and enterprises to get on board during conferences and events.

I Met People Through the MUSIC Project I Would Not Have Met Without It. That Is Very Positive This statement was made in regard to groups of both change agents and city officers. Furthermore, new contacts could be achieved with multiple departments whose area of function overlap with the topics of the guiding principles. The participants see the benefit of working together, and different departments are interested in collaborating with the actual working groups. The added value is seen in the exchange of expertise and knowledge. The synergies from these ties would be very supportive in mobilizing different kind of resources. For example, in the city context and the working groups, it seemed important to establish ties with the oil and gas industry as well as with the private sector because many activities in the city are building on these sectors but the arena process was not able to strengthen or keep contact with change agents from these businesses.

Operational Value of Envisioning: Vision as the Symbolic Space that Connects Long-Term and Short-Term Actions and Projects The commitment of the participants to realise the transition vision is achieved and expressed openly. This result occurred not only from the main change agents being already involved for more than 18 months but also because other actors were engaged throughout the process by the transition team that explicitly identified synergies between the new-coming actors and the ongoing process. All participants share the interest in the future of Aberdeen and Aberdeenshire that was maintained throughout the TM process and enabled bringing people together who are generally interested in sustainability issues.

The TM approach was mentioned as very effective for routing the dialogues on possibilities of Aberdeen's future as a sustainable city rather than only pointing out the failures and vulnerabilities. The change agents voiced that the created vision is not only motivating for oneself but increased motivation internally in the group.

The change agents see all guiding principles creating a 'future desire symbol' and in this way contributing to engaging action for achieving a future sustainable city. With the vision as a 'desire-capturing symbol for sustainability,' the change agents see it as their task to contact the missing actors and engage them to work together but claimed also that the general policy structure is not made for future-oriented thinking and that a shift on the decision-making level is needed to enhance the impact of the process.

The interviewed change agents did not experience missing actions as a demotivating factor. They highlighted the long-term aim of the process, pointing out that 3 years is not a long time to implement a shift in mindset as a precondition for carbon reductions. They are optimistic about their capabilities to contribute and are willing to identify and contact the actors who are yet missing to realise the upcoming projects. The possibility to contribute, however, does rely on the continuance of the working groups to build on that platform.

It also became apparent that the central role of the city council is to facilitate a follow-up process so the change agents proceed with the needed tasks for vision realisation. The change agents did not want to take up choices independently of public policy actions but, instead, they expressed the interest to seek for possibilities of their vision to overlap with the projects of the city and to create synergies rather than to fight the system, although the political structure was detected as hindering the ambition of the project

Reflexive Value of Envisioning: Learn to Recognise Connectivity and Engage in Politics of Space During the dialogues about future desires and aspirations about the city's sustainable future pathway, participants recognised that for achieving a sustainable future they need to collaborate in short-term actions and in keeping the track as well as recontextualising the long-term vision. In this context, it became evident that issues such as energy security and energy equality are interrelated, aspirations of inclusivity and environmental health together with pride are connected, and conflicts about these need to be openly arbitrated. Reflecting on these dialectic dynamics and learning, we argue that the envisioning arena became an arena where citizens learn about politics of space (Healey 2010, p. 9). In Aberdeen's context of envisioning, we also observed that:

> "People become aware of how their concerns inter-relate not only with those of their neighbors, but with those of people elsewhere whose concerns are raised in the discussion" (Healey 2010, p. 9).

References

Aberdeen City Council (2009) Scotland's climate change declaration report 2009

Aberdeen City Council (2010) Scotland's climate change declaration report 2010

Aberdeen City Council (2011) Behind the granite: Aberdeen key facts 2011

Beers PJ, Veldkamp A, Hermans F, van Apeldoorn D, Vervoort JM, Kok K (2010) Future sustainability and images. Futures 42:723–732

Berkhout F, Hertin J (2002) Foresight future scenarios: developing and applying a participative strategic planning tool. Greenleaf, Austin, USA
Bishop P, Hines A, Collins T (2007) The current state of scenario development: an overview of techniques. Foresight 9(1):5–25
Carlsson-Kanyama A, Dreborg KH, Moll HC, Padovan D (2008) Participative backcasting: a tool for involving stakeholders in local sustainability planning. Futures 40:34–46
Frantzeskaki N, Walker W (2013) Concepts and methods of policy analysis. In: Thissen WAH, Warren W (eds) Public policy analysis: new developments, international series in operations research and management science. Springer, Berlin, pp 261–272. ISBN 10: 1461446015
Frantzeskaki N, Loorbach D, Meadowcroft J (2012a) Governing transitions to sustainability: transition management as a governance approach towards pursuing sustainability. Int J Sustain Dev 15(½):19–36
Frantzeskaki N, Ferguson BC, Skinner R, Brown RR (2012b) Guidance manual: key steps for implementing a strategic planning process for transformative change. Dutch Research Institute for Transitions, Erasmus University Rotterdam. Monash water for liveability. Monash University, Melbourne. ISBN 978-1-921912-14-6
Frantzeskaki N, Wittmayer J, Loorbach D (2014a) The role of partnerships in 'realizing' urban sustainability in Rotterdam's city ports area, the Netherlands. J Clean Prod 65:406–417
Frantzeskaki N, Rogers B, Loorbach D, Brown R (2014b) Designing transition management processes. Principles and applications. Working Paper
Gidley JM, Fien J, Smith JA, Thomsen DC, Smith TF (2009) Participatory futures methods: towards adaptability and resilience in climate-vulnerable communities. Environ Policy Gov 19:427–440
Healey P (2010) Making better places, the planning project in the twenty-first century. Palgrave-Macmillan, New York
Helling A (1998) Collaborative visioning: proceed with caution!: results from evaluating Atlanta's vision 2020 project. J Am Plan Assoc 64(3):335–349
Hughes N (2013) Towards improving the relevance of scenarios for public policy questions: a proposed methodological framework for policy relevant low carbon scenarios. Technol Forecast Soc Chang 80:687–698
Keeney RL (1996a) Value-focused thinking: identifying decision opportunities and creating alternatives. Eur J Oper Res 92:537–549
Keeney RL (1996b) Value-focused thinking: a path to creative decision making: chapter 3. Harvard University Press, Cambridge, MA
Larsen K, Gunnarsson-Ostling U (2009) Climate change scenarios and citizen participation: mitigation and adaptation perspectives in constructing sustainable futures. Habit Int 33 (3):260–266
Larsen K, Gunnarsson-Ostling U, Westholm E (2011) Environmental scenarios and local-global level of community engagement: environmental justice, jams, institutions and innovation. Futures 43:413–423
Loorbach D (2010) Transition management for sustainable development: a prescriptive, complexity-based governance framework. Int J Policy Adm Inst 23(1):161–183
Nevens F, Frantzeskaki N, Loorbach D, Gorissen L (2013) Urban transition labs: co-creating transformative action for sustainable cities. J Clean Prod 50:111–122
Newman P, Jennings I (2008) Cities as sustainable ecosystems: principles and practices. Island Press, London
O'Brien FA, Meadows M (2013) Scenario orientation and use to support strategy development. Technol Forecast Soc Chang 80:643–656
Ozkaynak B, Rodriguez-Labajos B (2010) Multi-scale interaction in local scenario-building: a methodological framework. Futures 42(9):995–1006
Pichert D, Katsikopoulos KV (2008) Green defaults: information presentation and pro-environmental behavior. J Environ Psychol 28:63–73
Ringland G (2010) The role of scenarios in strategic foresight. Technol Forecast Soc Chang 77:1493–1498

Robinson J, Burch S, Talwar S, O'Shea M, Walsh M (2011) Envisioning sustainability: recent progress in the use of participatory backcasting approaches for sustainability research. Technol Forecast Soc Chang 78:756–768

Rogers T, Bazerman MH (2008) Future lock-in: future implementation increases selection of "should" choices. Organ Behav Hum Decis Process 106(1):1–20

Roorda C, Wittmayer J, van Steenbergen F, Frantzeskaki N, Loorbach D, Henneman P (2014) Transition management in the urban context: addressing climate change at the local level – Montreuil, Rotterdam, Ghent, Aberdeen. DRIFT, Erasmus University Rotterdam, Rotterdam/Ludwigsburg

Schultz PW, Nolan J, Cialdini RB, Goldstein NJ, Griskevicius V (2007) The constructive, destructive, and reconstructive power of social norms. Psychol Sci 18:429–434

Sheppard SRJ, Shaw A, Flanders D, Burch S, Wiek A, Carmichael J, Robinson J, Cohen S (2011) Future visioning of local climate change: a framework for community engagement and planning with scenarios and visualization. Futures 43:400–413

Shipley R (2000) The origin and development of vision and visioning in planning. Int Plan Stud 5:227–238

Shipley R, Newkirk R (1999) Vision and visioning: what do these terms really mean? Environ Plan B 26:573–591

Sondeijker S, Geurts J, Rotmans J, Tukker A (2006) Imagining sustainability: the added value of transition scenarios in transition management. Foresight 8(5):15–30

The Cities Alliance (2007) Liveable cities, the benefits of urban environmental planning, a cities alliance study on good practices and useful tools. Cities Alliance/ICLEI/UNEP, Washington, DC

Tompkins EM, Few R, Brown K (2008) Scenario-based stakeholder engagement: incorporating stakeholders preferences into coastal planning for climate change. J Environ Manag 88:1580–1592

Van der Helm R (2009) The vision phenomenon: towards a theoretical underpinning of visions of the future and the process of envisioning. Futures 41:96–104

Van Eijndhoven J, Frantzeskaki N, Loorbach D (2013) Connecting long and short-term via envisioning in transition arenas, how envisioning connects urban development and water issues in the city of Rotterdam, the Netherlands. In: Edelenbos J, Bressers N, Scholten P (eds) Connective capacity in water governance. Ashgate Publications, London

Volkery A, Ribeiro T (2009) Scenario planning in public policy: understanding use, impacts and the role of institutional context factors. Technol Forecast Soc Chang 76:1198–1207

Vural-Arslan T, Cahantimur A (2011) Revival of a traditional community engagement model for the sustainable future of a historical commercial district: Bursa/Turkey as a case. Futures 43:361–373

Wiek A, Binder C, Scholz RW (2006) Functions of scenarios in transition processes. Futures 38:740–766

Chapter 5
Interactions Among Multiple Niche-Innovations and Multi-regimes: The Case of the "Welfare Mall" in Higashiomi

Satoru Mizuguchi, Kyoko Ohta, P.J. Beers, Michiko Yamaguchi, and Toshiaki Nishimura

Abstract This chapter focusses on the dynamics of interactions among multiple niche-innovations and multi-regimes that occur under the pressure of slow but structural landscape changes. The energy transition, for example, is one of the core and broad changes that fundamentally impact not only the energy system but also systems of food, agriculture, and healthcare, as well as patterns of economic activity. Little work has been done on sustainability transitions that involve multiple niche-innovations, which may interact with each other as well as with regimes. This chapter investigates the case of the city of Higashiomi, Shiga, Japan, where a community business project called the "Welfare Mall" clustered local production of food, energy, and elderly care in one shopping mall-like venue. The project is ongoing as of this writing. This chapter focuses on the common struggles and challenges of regimes in different domains under the pressure of fundamental landscape shifts. Using the concepts of multi-level and multi-regime, the central issue concerns how the bottom-up attempt of multi-niche innovations from a geographically local context interacts with multiple regimes. We identify the multi-niche as a nexus of geographic proximity and multi-domain interaction, and therefore as a typical example of urban transition. Furthermore, the case underscores the relevance of reflexive activities for urban transition management.

S. Mizuguchi (✉)
Japanese Forum of Environmental Journalists, Tokyo, Japan
e-mail: msatoru99@gmail.com

K. Ohta
Policy Alternatives Research Institute, University of Tokyo, Tokyo, Japan

P.J. Beers
Dutch Research Institute for Transitions, Erasmus University Rotterdam, Rotterdam, The Netherlands

M. Yamaguchi
Environmental Department, Higashiomi City, Higashiomi City, Shiga, Japan

T. Nishimura
NOUGAKU Consulting.co, Shiga, Japan

D. Loorbach et al. (eds.), *Governance of Urban Sustainability Transitions*, Theory and Practice of Urban Sustainability Transitions,
DOI 10.1007/978-4-431-55426-4_5

Keywords Niche • Innovation • Multilevel perspective (MLP) • Multi-regime • Multi-niche • Learning • "Welfare Mall" • Scaling up • Transition management

5.1 Introduction

5.1.1 Scope

According to the multilevel perspective (Geels 2004; Geels 2011), slow but dynamic landscape changes influence the stability and sustainability of socio-technical regimes. The energy transition is one of the broad core changes that fundamentally impact not only the energy system but also systems of food, agriculture, and healthcare, as well as patterns of economic activity. The dominance of cheap and abundant fossil fuels has changed many existing regimes. Food production, for example, used to be small scale within local energy cycles (Altieri 1985), but has become large scale, monocultural, and highly intensive with the aid of gas-guzzling machinery. Since the mid-1970s, however, the availability of cheap oil has fluctuated. During this time, various innovative ways of farming emerged, such as organic farming. Local production of renewable energy also started to be used for local heating systems, food production, and consumption. In successful cases, these changes led to the creation of new jobs and improvement of local welfare.

The example of energy transitions becoming intertwined with food and agricultural transitions shows that transitions do not occur in isolation. Instead, societal systems in transition are also open systems that interact with other (changing) systems. Indeed, energy niches may come to impact agricultural regimes and the other way around. As Grin et al. (2010, p. 78) suggest, many transitions, especially sustainability transitions, involve multiple niche-innovations, which may interact with each other as well as with regimes. However, to date this topic has received little scientific treatment.

This chapter investigates the case of the city of Higashiomi, Shiga, Japan, where a community business project called the "Welfare Mall" clustered local production of food, energy, and elderly care in one shopping mall-like venue. The project is ongoing as of this writing. The Welfare Mall represents a clustering of multiple niches; in other words, it is a multi-niche. This chapter also focuses on the common struggles and challenges of regimes in different domains under the pressure of fundamental landscape shifts. Using the concepts of multi-level and multi-regime, the central issue concerns how the bottom-up attempt of multi-niche innovations from a geographically local context interacts with multiple regimes, and how this informs transition management in urban regions.

5.1.2 *The Multilevel Perspective*

Transitions takes place at multiple levels, and the multilevel perspective (MLP) has become commonplace for studying transitions and transition management. Grin et al. (2010) highlight the importance of the MLP for understanding sociotechnical transitions (Grin et al. 2010, pp. 18–9). Conceptually, the MLP distinguishes three levels: (1) technological niches (niche-innovations), (2) sociotechnical regimes, and (3) sociotechnical landscape (Fig. 5.1). The MLP is a middle-range theory that sees each level as being nested in (organisational) hierarchy.

Much existing MLP literature is primarily concerned with the relations and interactions between niches and regimes (Kemp et al. 1998; Geels 2004, 2011; Schot et al. 1994; Schot and Geels 2008). The regime, on the one hand, is characterised by stability and lock-in mechanisms. According to Geels (2004), regimes contain three types of rules: cognitive, regulative, and normative. The niches, on the other hand, harbour innovations (novelties) that deviate radically from the regime.

The core notion of MLP is that transitions come about through interactions between processes at different levels: (a) niche-innovations build up internal momentum; (b) changes at the landscape level create pressure on the regime; and (c) destabilization of the regime creates windows of opportunity for niche-innovations. Niche-innovations are still important, but they can only diffuse more widely if they link up with ongoing processes at regime and landscape levels (Grin et al. 2010, p. 88). To foster this process, transition management scholars such as Loorbach (2007, 2010) argue that a combination of multilevel activities is necessary, some strategic and long term, some tactical and mid-range, and some operational,

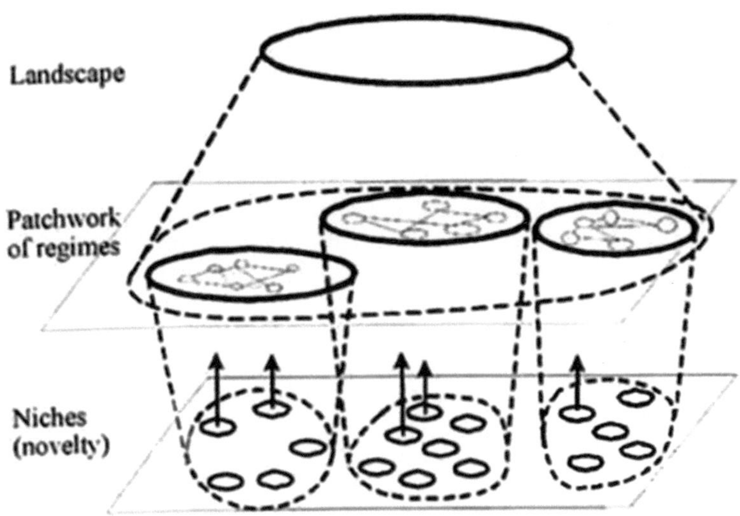

Fig. 5.1 Multiple levels as a nested hierarchy (Geels 2002, p. 1261, from Grin et al. 2010, p. 19)

putting a stake in the ground. And, perhaps most importantly, reflexive activities are needed to connect ongoing development (strategic, tactical, operational) to the continuously changing state of the system and the regime.

Both the governance approaches of transition management (Loorbach 2007, 2010) and strategic niche management (SNM; Kemp et al. 1998) emphasise the importance of novelties in niches. Neither approach suggests that governments create niches in a top-down fashion. Rather, they focus on endogenous steering, steering from within. Such steering can address many parts of the process, by adding a new actor, a specific learning process, or a set of demonstration projects that may redirect evolving dynamics towards a desired path (Grin et al. 2010; Rip 2006; Nill and Kemp 2009). In so doing, a process of *experimentation* aims to foster the transition potential of niche novelties.

In SNM, the analytical focus is mainly on the micro-level and niche actors' strategic behaviours (Vernay et al. 2014). Transition management adds the *transition arena* as a specific governance space. The arena consists of frontrunners that are able to connect ongoing experimentation with future visions of changed regimes and an awareness of landscape developments. Landscape pressures do not mechanically influence regimes. Instead, this influence is mediated by actors' perceptions, negotiations, and agenda setting (Grin et al. 2010, pp. 25–26). This definition underscores the importance of learning/monitoring, not only at the niche level but also in transition arenas. Therefore, the first question in this chapter concerns how local niche-innovations have emerged regionally, and through which learning processes.

The second question is related to the first one, but is more about the spatial/geographic aspects of niche-innovation (Coenen et al. 2012). In some cases, synchronicity between niches occurs: various niches of different domains emerge at the same time and at the same place/local area (Coenen et al. 2012). When multiple niches are intertwined and can strengthen each other, transition may be more likely to occur, although the different niches may also come to hinder each other. Therefore, our second question is what enables positive interactions among multiple niches within a geographic region.

The third argument of this chapter is about scaling up. The higher the scale level, the more aggregated the components and the relationships and the slower the dynamics among these actors, trends, and developments. Scaling up is to influence the system's dominant structure, culture, and practice. Against these bottom-up challenges, regime actors have reactions in three different ways: (1) defensive approach, (2) reactive and accommodating approach, and (3) innovative approach (Grin et al. 2010, p. 133). Scaling up refers to embedding a transition experiment in established ways of thinking (culture), doing (practices), and organising (structure). Through scaling up, a new or deviant constellation of culture, practice, and structure attains higher influence and stability and increases its share in meeting a societal need. Scaling up implies that sustainable practices that are initially unusual have become the dominant or mainstream practice (Grin et al. 2010, p. 209; Loorbach and Rotmans 2006).

If there are interrelated multiple niche-innovations of different domains, there should be related multiple-regimes (Raven and Verbong 2007; Geels 2007). Loorbach (2010) suggests there is a "patchwork of regimes," rather than a single regime, in which (sub-)transitions take place at different speeds and at different moments in time. Hence, the third and the final question: what is the process of scaling up of transition through multiple niche-innovations and multi-regimes?

In Chap. 2 in this volume, Wittmayer and Loorbach discuss the specifics of transitions in urban geographic systems in terms of geographic proximity, policy proximity, personal proximity, multi-domain interaction, and multi-scalar interaction. In our discussion of the results we also discuss the implications for these urban transition specifics.

5.2 The Case and Methodology

5.2.1 Case: The Higashiomi "Welfare Mall"

The preliminary work for the "Welfare Mall" project started May 2009, and the process of materializing the plan through various discussion arenas happened in 2010–2012. It then moved into the operational stage in April 2013, when the niche-innovative facilities attracted interest beyond the local community.

The mall's visible parts consist of a slow-food restaurant, an elderly-care facility, a bakery café with wood-for-fuel workshops, and solar photovoltaic cells on all the rooftops. The staff includes elderly persons, handicapped persons, and young persons. For them, it is a place to work and learn.

Invisible as this seems, the mall links up with social innovation networks in related fields. The slow-food restaurant is based on groups of small farmers, senior housewives willing to convey their knowledge of local cuisine, and elderly customers of catering services. The elderly-care facility is supported by a voluntary network of medical doctors, caregivers, home helpers, nurses, monks, citizens, and the library. The bakery café with its wood-for-fuel workshop is the product of collaboration among citizens, business, and the municipality. As described later, their collaboration reduced wood production costs to a level that beats fossil fuel costs. The solar panels belong to the Citizen's Electric Power Cooperative, with more than 300 investors who receive profits from electricity sales through a local currency. The local residents' hope for living in their community with a sense of security, even when they become old and frail, is embodied in the Welfare Mall. Its associated networks serve as an alternative model for the welfare state under pressure.

As the production of food and fuel as well as the delivery of healthcare services are all conditioned by the post-industrial and globalised environment, the struggle to keep traditional ways of production and delivery in path-dependent and sectionalist institutions has faced its limitations. At the same time, however, countercurrents can be observed since the rise of the environmental movement in the 1970s and climate change issues in the 1990s, two events that constitute the main landscape changes for environmental as well as energy, agriculture, and healthcare systems. Nowadays, locally produced food and renewable energy are increasing in their market shares. National healthcare policies are also being revised from being hospital centered to being home- and community centered, partly encouraged by the national Long-Term Care Insurance introduced in 2000. The Welfare Mall, the cluster of providers of local food, energy, and elderly care, in that sense reveals a possible direction for a more sustainable society.

After reviewing the context of Higashiomi, the following sections investigate novelties in the fields of food, energy, and elderly care in the area, investigating how each developed, how it interacted with the regime, and under what landscape changes. This chapter then describes how local novelties linked up in the form of a multi-niche.

5.2.2 Higashiomi's Geographic Context

Higashiomi means the east (Higashi) of the Big Lake (Omi). It is a municipality with a population of 120,000, resulting from mergers of seven smaller municipalities during 2005–2006. Its population as well as geographic size amount to 1/1000th of Japan. Its land use is also similar to that of all of Japan; that is, 65 % is forest, 14 % farmland, and the rest of the land use is composed of industrial, commercial, and residential areas.

At 40 km to the west, across mountainous terrain, lies Kyoto, the ancient capital of Japan, and 60 km east, also across a mountainous area, lies Nagoya, the industrial capital of Japan, the home city of Toyota Motor Corp.

With Japan's largest lake nearby, an abundant water reservoir, and its proximity to Kyoto-Osaka-Kobe and Nagoya, two of Japan's big industrial areas, large companies have located their manufacturing facilities in Higashiomi. Many residents work for local factories and offices on weekdays while practising small-scale farming during the weekends.

5.2.3 Methodology

The following case study is based on a combination of the following methods.

1. Fieldwork and documentary research to clarify trajectories of local practises at niches on food, energy, and healthcare as well as a map of frontrunners and gatherings.
2. In-depth interviews for clarifying the Welfare Mall's funding scheme.

We conducted in-depth interviews in Higashiomi city and analyzed official as well as NPO papers, written materials, and websites of both the local government and other public and private organisations. Most interviews were carried out on July 18–19, 2013 and on February 10, 2014, with more than ten key persons, including officials of the municipality (first-tier government, some of which also work with local NPOs), senior and frontline staff representatives of NPOs (including one ex-official), a senior official of the Shiga Prefecture (second-tier government), and various community entrepreneurs. Some of the interviewees performed multiple roles, for example, local official, NPO staff, community entrepreneur, and/or representative of central and local government committees. We tried to be sensitive about how an interviewee's comments and perspectives were informed by their various roles, while at the same time keeping the uniqueness of the ways of thinking and behavior of these "multiple-role" interviewees.

5.2.4 *Analysis*

Analysis was focused on interactions among key persons in multi-niches, and on how these interactions contributed to the sequence of events leading to another novelty. Grin et al. (2010, p. 94) argue that "in process theory the world is made up of entities that participate in events and may change their identity" and "for process theory, central subjects are individual entities (people, group, etc)." This chapter analyses the different niches first by looking at main actors, the practise in the niches, as well as the physical infrastructure and how the niche compares to the regime. For each niche, the different novelties were identified and then the interconnections were analysed.

5.3 Multi-niche-Innovation and the Welfare Mall

5.3.1 *The Societal Landscape*

The local "landscape" has changed from the middle of the 1990s after the collapse of the economic bubble. Along with the high appreciation of the yen and deflationary policies, some local factories went bankrupt, and others moved overseas. The possibility of another factory exodus sparked anxiety among local residents. Farming income also began to drop because of the price of rice. This Japanese staple food gradually decreased as it was faced with foreign competition, significant reduction of farm subsidies, and declining demand because of the ageing of the population.

National government revenue deficits also brought healthcare budget cuts for local governments. Some public hospitals needed to close. Elderly care was also targeted by budget cuts. A sense of insecurity spread among senior residents. The younger generation was a victim of the sluggish economy as well, having difficulties getting full-time jobs even at the highest levels of education. The fieldwork and interview data suggest that a mounting sense of insecurity about community life gave residents cause to rethink their community resources.

5.3.2 Local Practices at Niche on Food

The slow-food restaurant in the Welfare Mall has inherited several novelties from other local non-profit organisations.

The first novelty is the menu, based on more than 200 recipes collected from local families. The menu thus is the reflection of ordinary local family foods, ranging from traditional to modern. It was the fruit of the "Discover Family Food" campaign during 2009, with a series of sampling parties where hundreds of residents have participated with their meals and recipes. The organiser, the Council of Hand-Shaking Higashiomi was founded in 2008 by five Non-Profit Organisations (NPOs), four community organisations, and the municipality. Their purpose was "finding and organising" untapped local resources for community development. Through the campaign, various "endangered" local recipes were not only saved but also inherited by younger generations. Some of the participants joined the slow-food restaurant in the mall as chefs and staff.

The second is diversity of local produce: more than 200 varieties of vegetables were supplied by small farmers. Some of the farmers resumed growing obsolete species.

The third is the catering service for the elderly and local business persons, who are unable to cook for themselves but are willing to eat locally produced and prepared meals. Catering is conducted by mentally handicapped people. The job also means "empowering" the mentally handicapped for more advanced job placement through conversing with "others." This program is subsidised by the Ministry of Health, Labour and Welfare.

The fourth aspect is the restaurant, equipped with free spaces for meeting, chatting with neighbours, and holding seminars, the themes of which range from how to raise a child, to how to cook local food, et cetera.

The fifth is that the restaurant is also a workplace for the elderly and the mentally handicapped.

The core organisation for the council is the Rape Blossoms Eco Plaza of Aito (Aito was a smaller municipality before the merger), dating back to 1977 when activists protested against pollution of the largest lake. Most of them were housewives who had been lobbying to ban the sale of phosphorus-containing synthetic detergent, one of the causes of pollution. They had also produced a biodegradable detergent by themselves from used cooking oil, which was another source of pollution. The citizens had been organising a closed loop for collecting used cooking oil as a resource for soap, fuels, and fertiliser during the 1980s. Since the 1990s they have also been growing rapeseed to meet growing local demands for fuels and fertiliser. Through the process, Ayako Fujii, a leader of the movement, has been appointed as governmental advisory board member for the Ministry of Agriculture, Forestry and Fishery as well as the Environmental Agency since the 1990s. Based on her local practices as well as foreign knowledge, she has been actively lobbying for sustainability issues. Her policy recommendation activities to central government sometimes lead to acquiring subsidies for her citizens' organisation. The Rape Blossoms Eco Plaza of Aito and the "Discover Family Food" campaign were partially government subsidised.

Japan's mainstream agriculture policy has been promoting monoculture (rice and several selected vegetables) tied with production-stimulating subsidies for larger farmers while remaining dependent on imports of feed, fuel, and fertiliser. Promoting small farmers, obsolete local varieties, slow food, and local production of fuel and fertiliser through resource-circular farming were absent from the regime until the turn of the twenty-first century. Issues of environment, climate change, and the need for local revitalization under a budget deficit opened windows of opportunities for niche developments such as the Rape Blossoms Eco Plaza of Aito.

Another regime shift came from a tightening of the welfare budget. Turning the handicapped into taxpayers (from tax recipients) became an urgent agenda for the welfare regime, and for this reason catering by the handicapped was partially subsidised by the Ministry of Health, Labour and Welfare.

In sum, the slow-food restaurant in the Welfare Mall was a bricolage of various developments and novelties—the diversity of menu based on the diversity of local produce, including obsolete varieties, connecting the agricultural and welfare regimes through job creation for mentally handicapped people as well as the elderly, leading to their empowerment and tapping their potential—all of which were developed in niches and partly subsidised by multi-regimes, including those of agriculture and welfare.

5.3.3 *Local Practises at Niches on Sustainable Energy Production*

The bakery café with its wood-for-fuel workshop as well as the Citizens' Electric Power Cooperative in the Mall (combining forestry, energy, and economic niches) both have also inherited several novelties from other local developments in which innovative energy production systems incubated.

The first is an initial funding scheme for Citizens' Electric Power Cooperative. Its origin dates back to 2003, when the Renewables Portfolio Standard (RPS) Act was enacted. For the first time in Japan, the Act had obliged electric power companies to buy electricity from suppliers of energy from renewable sources. The Citizens' Council for Promoting New Energy at Yohkaichi (another smaller municipality before the Higashiomi merger) had organised a series of meetings to start the first Citizens' Electric Power Project in 2004 (Nishimura and Nomura

2013). The idea of initial funding through mobilising citizens' investments was contributed by a knowledgeable "outsider" from another region, who was already involved in a similar project elsewhere. Another outsider was Toshiaki Nishimura, a founding member of the Council. An engineering consultant, he moved from Kyoto to Higashiomi for a better environment for his children.

The idea of introducing a local currency as a means for returning sales profits to the investors was born in 2009, when the second Citizens' Electric Power Project was under development. The local currency benefitted the local economy and it gave the Citizens' Council opportunities to "organise" local business people. The concept of local currency itself had been already popularised by a change-minded central bureaucrat from the Ministry of Economy, Trade and Industry around 2002.

A series of preparatory meetings for the second project were partly subsidised by the Ministry of Environment as community environmental work.

Along with these preceding practises, the Welfare Mall was able to use a Feed-in Tariff (FIT) scheme, which was introduced in 2012 by the Agency for Natural Resources and Energy. The combination of initial funding through mobilising citizens' investment and introducing a local currency as a means for returning sales profits to investors originated in Higashiomi as the "Higashiomi model"—a combination of an economic niche and an energy niche—appraised by the Minister of Internal Affairs and Communications (Ministry of Internal Affairs and Communications 2011). The introduction of the FIT by the government helped to facilitate initial funding and secure long-term profitability for the third Citizens' Electric Power cooperatives installed in the Welfare Mall.

In sum, the Citizens' Electric Power Cooperative in the Welfare Mall was a bricolage of novelties—a combination of local production of renewable electricity, citizens' funding, and local currency, which together fostered the production of energy from renewable resources and the local economy.

The café with the wood-for-fuel workshop also "inherited" several local practises from other organisations, including the Council for Promoting Higashiomi Community business.

Fuel wood production was expensive compared with heat production from fossil fuels, which was supported by the government. However, the Council for Promoting Higashiomi Community business discovered that collaboration among citizens, local businesses, and the municipality can provide a framework to compete with heat production from fossil fuels (Nishimura and Yamaguchi 2011). The research was funded by the Ministry of General Affairs through the municipality. The research made clear that fuel wood production costs consist of harvest (51 %), production (30 %), and delivery (19 %). It also revealed options to reduce these costs. For example, organising local citizens for harvesting as part of recreation and an environmental education program have already been put into practise to reduce harvesting costs.

Another novelty was developed by Lovely Town Eco-club at Higashiomi, a non-profit organisation, which discovered that the use of new governmental subsidies in a combined way may also effectively reduce production costs. The research was funded by the philanthropic foundation of JT, Japan Tabaco. One subsidy was the sustainable forestry scheme by the Forestry Agency, subsidizing community forestry planning. The second was a governmental subsidy for population control of wild animals including deer, monkey, and wild boars to reduce damage to agricultural production. Without a maintained countryside forest as a buffer, wild animals tend to visit and damage farmland in search for food. The money came from the Ministry of Agriculture, Forestry and Fishery. The third subsidy was for empowering mentally handicapped people for more advanced jobs, from the Ministry of Health, Labour and Welfare. Handicapped people turned out to gain confidence and skill for doing teamwork through wood production. Mitsuko Nonomura, president of the Centre for Empowering Handicapped, a non-profit organisation, creatively used the subsidy, originally purposed for labour policy, for fuel wood production. She has been pioneering new job opportunities for the handicapped.

In both cases, citizen participation for community forestry constituted a vital part of wood production because much of the forestry labour depended on citizens' voluntary activities. People in Higashiomi had already experienced this in the late 1990s, when a degraded countryside community forest became a dumping site, which consequently led wild animals to come and visit nearby farmland. High appreciation of the yen, free trade of timber, and cheap fossil fuels since the 1970s made domestic production of timber and wood less attractive for forest owners, which made them indifferent to their forest. Abandoned forests attracted illegal waste disposal as well as wildlife, which led to further forest degradation during the 1980s and 1990s.

In the middle of the 1990s, several people volunteered for cleaning up the forest because they had good childhood memories of playing there. They trimmed bush, collected garbage, cut a footpath, and in so doing turned the forest into a more comfortable place for walking, playing, and learning about nature. One of them was Seizo Muto, a senior official of the municipality, whose specialty was financing. He devised using the municipality's waste treatment budget for handling garbage in the forest. Currently, a nature education centre is located in the forest, hiring three full-time staff (Mizuguchi 2013).

In sum, the fuel wood workshop in the Welfare Mall was a bricolage of novelties—connecting the forestry, waste, and welfare regimes on the one hand, and aligning voluntary and business sectors on the other. Both developed in niches through citizen activism partly supported by a philanthropic foundation, the municipality, and central government funding.

The Citizen's Electric Power Cooperatives and fuel wood production have evolved independently. However, they have shared key persons who belong to the same "learning organisation," including the Lovely Town Eco-club at Higashiomi.

5.3.4 Novelties in Elderly Care

The elderly-care facility in the Welfare Mall has also inherited several novelties from other organisations, including the Citizens' Roundtable for Thinking Medical Care and Welfare from Community Point of View, which we describe later. The facility aimed to support elderly care at home, even in cases of dementia or disability.

The novelties are supported by a network of medical doctors in various specializations, care workers, nurses, home helpers, monks, and citizens. Japan's medical care system has been divided by pursuing specialization without a coordinating scheme. Consulting with a general practitioner has become rare among Japanese since the growth of the hospital industry in the 1970s.

Elderly-care systems were also divided and complicated. Traditionally, government has been relying on elderly care by family members (especially housewives) and mutual help among neighbours, with only residual care by the government's welfare placement system (*sochi* system). During the high economic performance of the 1960s and 1970s, the government's elderly-care policy had been a kind of "trial and error" against the backdrop of optimistic expectations about funding during economic growth and the generous ruling party, looking at the elections. The symbolic policy (error) was free healthcare for the elderly (65 and older) in the 1970s, which was later abolished. As elderly care by family and neighbours was no longer feasible in many parts of Japan after the 1990s, the Ministry of Health, Labour and Welfare started discussions and experimental programs with select municipalities to introduce the national Long-Term Care Insurance,[1] which marked a turning point from elderly care by family to elderly care by a professional facility. Higashiomi was among the discussion groups. Kenji Kitagawa, a senior municipality official, moved from the prefectural government and was one of those who were able to converse directly with bureaucrats at the welfare regime.

The Long-Term Care Insurance has been contributing to the growth of the elderly-care industry in accordance with the growing elderly population and their expanding needs. However, coordination between elderly care and medical care was insufficient. To make matters worse, government revenue deficits resulting from the deflationary policy burdened welfare budgets. Proponents of the elderly-care regime again turned their eyes to community healthcare (residential care and homecare by small-size businesses and voluntary neighbours as well as families). They initiated a discussion to shift from expensive, private care homes to residential care. The novelties in elderly and medical care in the Higashiomi niches had attracted their attention.

[1] In 2000, the government introduced the Long-Term Care Insurance system, a social insurance system aimed at the rapid development of various types of care provision with the help of market forces. Considered by many as a step in the direction towards a *socialization of care*, high expectations mounted (Saito et al. 2010, p. 68).

In May 2009 the preliminary body for the Welfare Mall project was organised by "frontrunners" from elderly care, medical care, and the environment. Seizo Ohta, a community business owner for elderly care, was one of them. He had observed gaps among existing elderly-care policies and schemes. His dream was to help elderly persons gain a sense of security even in cases of dementia or disability (Lovely Town Eco Club 2012). Another key figure was Masaji Nomura, a senior municipality official and board member of the Rape Blossom Eco Plaza of Aito. He had thought about how to integrate policies across the environment, food, and elderly care for community revitalization. Takeshi Kokaji, architect and owner of both a private high school and a facility for the elderly, was another key person. He had already organised the Citizens' Roundtable for Thinking Medical Care and Welfare from a Community Point of View since 2007, which consisted of medical doctors, nurses, caretakers, monks, and citizens. The Citizens' Roundtable met every few months to discuss how to cooperate among various organisations in the medical and elderly-care system. Their aim was to realise a "seamless" web of care in a less expensive way. They had been creating face-to-face relationships between different medical and elderly-care organisations, and they were able to share patients' medical data amongst each other. This work served as a foundation for a seamless web of elderly care in the community.

The main novelties here were (1) building face-to-face, non-hierarchical relationships locally across highly hierarchical "kingdoms" of medical care and elderly care; and (2) sharing patients' medical data, as neither digitalization nor sharing of patients' data among medical and elderly-care institutions was in progress in Japan at that time. The central government had much interest in these developments in Higashiomi because they were eager to encourage voluntary cooperation among medical and elderly-care institutions as well as elderly care at home, which they thought would mitigate the growing burdens of medical and elderly-care cost. Thus, the government decided to subsidise the hardware of the Welfare Mall in 2012.

5.3.5 *The Welfare Mall: A Multi-niche of Food, Energy, and Care*

The novelties of food, energy, and care developed independently before 2008. However, a growing awareness grew among several key persons that they needed to cooperate more to make further steps towards a sustainable community. For that purpose, organising tools including a *frontrunner map* were developed.

At the same time several incidents (landscape changes) provided a window of opportunity for cooperation and alignment among key persons in multi-niches. The first was a future visioning process of a low-carbon municipality, the second was that the central government increasingly favoured local autonomy, and the third was a revision of the Long-Term Care Insurance. These three incidents provided a catalyst for the alignment among novelties in the form of a multi-niche.

One of the key actors was Michiko Yamaguchi, a middle-rank municipality official with a background in forest engineering and work experience in prefectural government after majoring in forestry at college. Incidentally, it was in 2008 that the science foundation of the Ministry of Education, Culture, Sports, Science and Technology selected the Higashiomi municipality to develop a municipal Low-Carbon Vision for the year 2030. The foundation had started a project about the scientific basis for realizing a low-carbon society against the backdrop of a shifting trend in the environmental policy regime. In January 2008, the then prime minister announced the national goal to achieve 50 % CO_2 reduction by 2050. However, local knowledge and skills were absent.

Higashiomi city and the Shiga Prefecture environmental research institute worked together to make a vision and a roadmap for halving CO_2 emissions by 2030 through innovating the local economy. Along with socioeconomic analysis and simulation, they organised a series of discussion meetings named the Environmental Roundtable for Low Carbon Vision. In these discussions, the methods of backcasting and scenario development were applied (Naito 2012). For organising the roundtable, Yamaguchi and her team thought they needed to involve not only key persons in the environmental policy area but also those in other areas including agriculture, forestry, commerce, industry, welfare, medical care, and urban planning. They thought it necessary to make a regional "frontrunner map."

Yamaguchi had already prepared a frontrunner map herself for organising another environmental project. Although the map listed only frontrunners in the environmental field, it provided a starting point for the new one. People around her helped her to make the map more comprehensive (Figs. 5.2, 5.3), also covering areas of sustainable development. Their criterion was to select frontrunners who are "independent, willing to work with other frontrunners in different fields and positive thinking," as described in the map.

Another incident was a change in the central government when the long-standing Liberal Democratic Party lost the general elections in August 2008. The new ruling party, the Democratic Party of Japan, which lasted until 2012, promoted local autonomy, locally produced renewable energy and restoration, and tried to collect associated good practises. Hisao Nishizawa, the then mayor of Higashiomi, and Yamaguchi were appointed as governmental committee members to collect good practises among municipalities. Higashiomi was regarded as a "good practise model" because of novelties such as the Citizens' Electric Power Cooperative, the Rape Blossom Project, the Citizens' Roundtable for Thinking Medical Care and Welfare from a Community Point of View, etc. The city was, in a sense, selected as "good practise city" network.

The municipality set up Green and Local Autonomy section. Masaji Nomura became its chief and Yamaguchi became a staff member, assigned to find and organise local frontrunners, which she had already been doing informally.

Since 2010–2011, the Round Table for Low-Carbon Vision met 20 times and provided opportunities for local frontrunners to engage with new points of view about the low-carbon municipality and sustainability. During the same period, more casual discussion "gatherings" were also organised for frontrunners, based on the

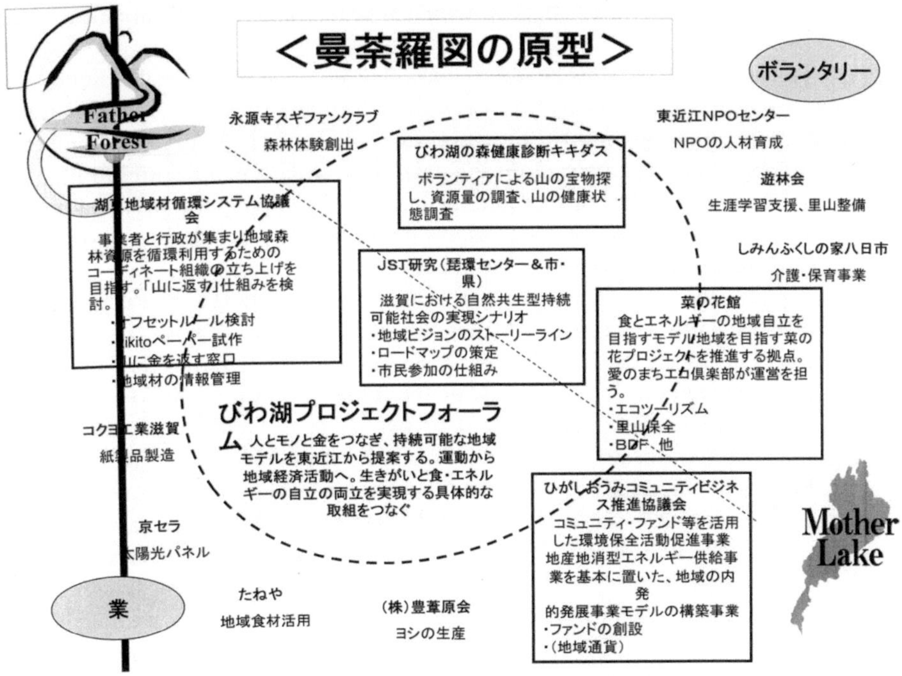

Fig. 5.2 First version of frontrunner map of Higashiomi in the environmental field

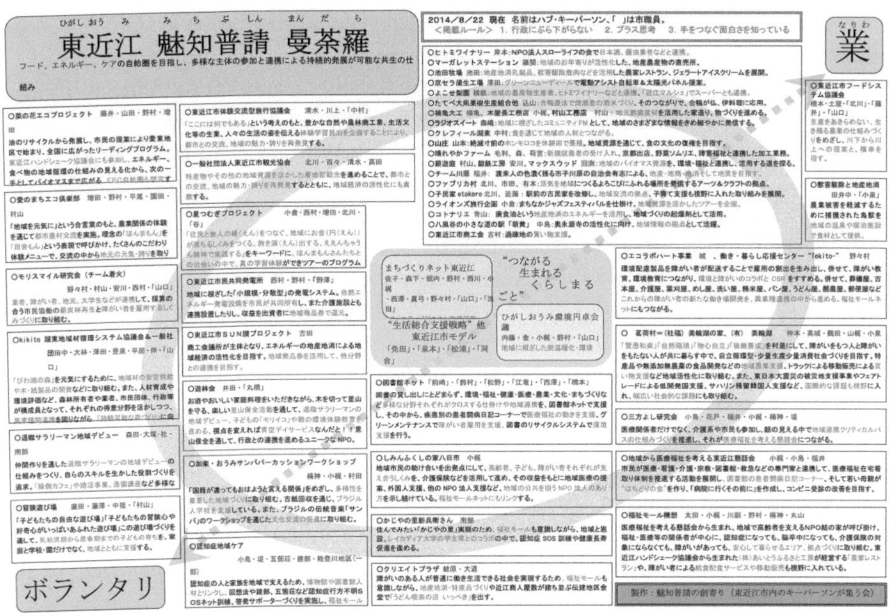

Fig. 5.3 Second version of frontrunner map of Higashiomi for local self-reliance for food, energy, and care

frontrunner map. Bringing the map on their hand, key figures with novelties met across sectors there.

More and more key persons began to think that organising the multi-niche was inevitable for further development for the region during 2007–2009. They founded a management committee in 2009 to realise the Welfare Mall with nine members: two from each of the three areas of disabled persons, elderly people, and food, plus three external advisors. They also organised a council in which various stakeholders and citizens could participate in discussions about the Welfare Mall, including the expansion of the project beyond the Aito area.

At this point, the individual services of the Welfare Mall were decided: job training and a bakery workplace for mentally handicapped people, daycare and short stay for the elderly, and a local supply-and-consume restaurant for food. This plan included a meal delivery service in the job training centre for the disabled, for the elderly in the daycare centre, and their own homes. Before the start of the Welfare Mall, each of these services had already been piloted for 3–4 years. In the Welfare Mall, a managerial board for the linked services and buildings was added. The integration and arrangement of three different facilities in one place made it possible to have, besides their individual services, linked and flexible services that could respond to needs that previously had fallen in between existing schemes.

Funding was the biggest issue for the project's success. At this point, the people in the network, especially those who had links with the central government, informed the bureaucrats who were in search for a model recipient of financial support in the fields of energy and welfare about the Welfare Mall. In this case, the Bureau for the Elderly in the Ministry of Health, Labour and Welfare, being in charge of the national Long-Term Care Insurance and looking for innovative model cases that could complement financially strained local insurance systems in more autonomous ways, responded to an approach by an officer of Shiga Prefecture.

The Welfare Mall project intended to apply for the Bureau for the Elderly's grant program for the encouragement of local arrangement of welfare space. The application procedure was via the municipality (in the Welfare Mall case, Higashiomi city) to the Bureau. The program's conditions had previously been limited to projects with services that were formally written in the care services arrangement plan of their municipalities. According to our interview, the Bureau apparently decided to change those conditions when they received the request for the Welfare Mall. In that year, it added a new category called 'project proposal grant,' which allowed applicants to use the grant more flexibly and autonomously, while still in collaboration with the municipality and under scrutiny by the donor in terms of innovativeness and feasibility. The fact that several officials of Higashiomi city were deeply involved in the planning was an advantage for the Welfare Mall application. The grant for the Welfare Mall was formalised in 2012 and the Mall was opened in the spring of 2013, followed by a number of visits by representatives from other municipalities as well as central government officials. Although the Welfare Mall itself may just as well have started without the frontrunner map or all the stakeholder gatherings, these tools did provide important networking opportunities to combine multiple niches and gave a context for future Welfare Malls, planning for which is under way as of this writing.

5.4 Reflections: Interactions Among Multiple Niche-Innovations and Multi-regimes

Transitions are long-term processes (40–50 years), whereas breakthroughs may be relatively fast (e.g., 10 years) (Grin et al. 2010, p. 11). It took 5 years for the Welfare Mall from initial meetings to opening. However, the predevelopment of the associated novelties took more than 10 years, and 40 years had passed from anti-pollution activism leading to community business based on local resources. In this sense, it is safe to say that the mall constitutes part of several transitions in and around Higashiomi. Based on the image of hierarchy processes in MLP (Fig. 5.1), this chapter has recognised the modified image of multilevel interaction observed in the Higashiomi case (Fig. 5.4).

At the landscape level, the three domains of food, energy, and elderly care have all exerted pressure on corresponding regimes. At the regime level, a "patchwork of regimes" is under both landscape and niche pressures. At the niche level, multiple niches of various and different domains emerge. At this, niches are combined and exert increasing pressure on existing regimes. We see this as the emergence of a

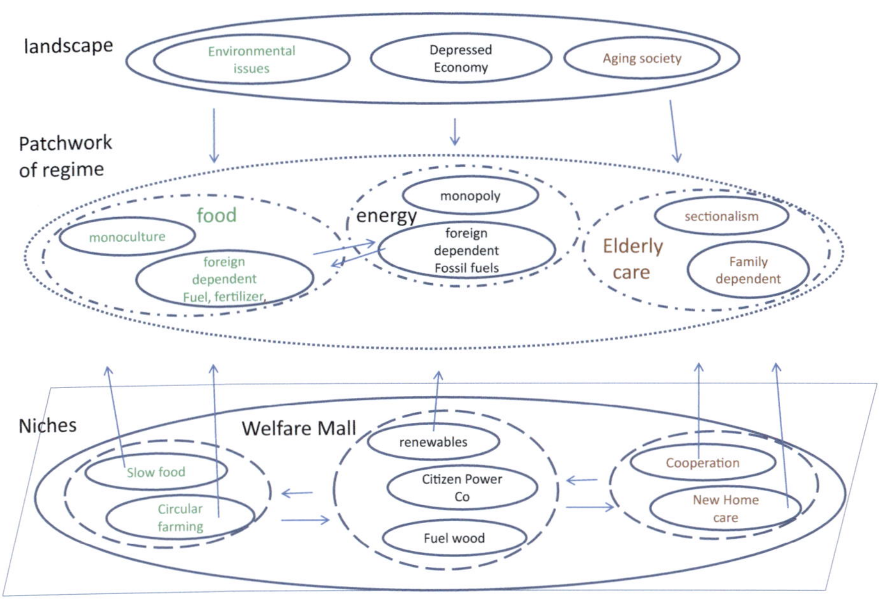

Fig. 5.4 Multi-level analysis of multiple niches-innovations and multi-regimes in Higashiomi (Mizuguchi S., design by Hosoi R.)

multi-niche, the physical manifestation of multi-domain interaction combined with geographic proximity (Wittmayer and Loorbach 2016, Chap. 2, this volume). In the multi-niche, interactions emerge with regimes of different domains, and each regime not only influences the others but also depends on it. Therefore, these regimes mutually stabilise each other for now, probably as a patchwork, until a more fundamental breakthrough occurs because of innovative multi-niche combinations.

The first question of this chapter was how local niche-innovations have emerged regionally and through which learning processes. The case study suggests that the following combination of elements had impacts in Higashiomi. First, there are combinations of actors including knowledgeable policy entrepreneurs at the niche level, and change-minded "outsiders" both within the municipality as well as in the central government. Frontrunners were willing to work with different fields. It is interesting that most of them belong to *learning networks* such as the Council of Hand-Shaking Higashiomi and the Rape Blossoms Eco Plaza of Aito. In these networks, participants were able to learn what is happening in the community, in Japan, and in the world through researchers, practitioners, bureaucrats, and each other. Furthermore, these networks were able to strategically secure central funding from across the ministries to support active experimentation. In so doing, an iterative learning cycle of action and reflection was pursued that is similar to the role of experimentation and learning/monitoring in transition management, and these networks are therefore reminiscent of transition arenas.

Additionally, several stakeholders held multiple jobs/societal roles: (1) a main job for basic income, such as government officer, NPO staff member, or entrepreneur; (2) a second job for the "change" of society and present situation, as NPO staff member, activist, or community entrepreneur; (3) a geographically embedded job, as farmer, forester, or volunteer and community staff at local temples, shrines, or neighbourhood association. These multiple societal roles may have enabled various key players to be able to bridge multiple perspectives and connect different values because they had multiple perspectives themselves.

The niche-level actors in Higashiomi tended to solve social issues through creating "community business models," using subsidies as seed money for creating viable business operations. This tendency is reflected in local traditions. Higashiomi is the birthplace of the so-called "Omi merchant," often referred to as the prototype of the modern Japanese merchant, whose pragmatic philosophy is known as "do well for seller, well for buyer, and well for society." People highly value both earning their bread and contributing to the local community. Here, the importance of transition management for creating experiments that bridge multiple values is clearly shown. If such people gather in a transition arena, the networking and collaboration would be much easier and encouraged, in line with the Regeer et al. (2011) concept of "Connected Value Development."

The second question was what enables the multi-niches positive interactions within the geographic region. Some of the concepts developed by evolutionary theories and innovation studies suggest that radical innovations often emerge outside or at the fringes of existing regimes, where niches act as incubation spaces that protect novelties against mainstream market selection (Grin et al. 2010, p. 22). Geographically, Higashiomi is located between Kyoto and Nagoya, which means Higashiomi is located on the fringes of two big socioeconomic and cultural centres. Neither had dominated Higashiomi. In this sense, a geographic *remoteness* from the regimes may have offered protective space.

The third and the final question concerned the process of scaling up a transition through a multi-niche and multiple regimes. The case shows that the use of central funding, particularly of project proposal grants, played a pivotal role in scaling up. An innovative project tried to change, through the "co-arrangement" of the government grant program, the mindset and behaviour of bureaucracy in municipalities as well as the upper tier of government. Indeed, a cross-disciplinary flexible grant program emerged within the central government, and there is a possibility that the arrangement encourages more policy innovation. According to this study's research, a limited number of central government officers explored how to use their budgets to encourage niches and innovation by municipalities and NPOs. In so doing, they acted in line with their own values and consciousness more than because of their departments' interests. This action provided an opportunity to niche actors to develop personal connections with these bureaucrats to inform them of their existence, to advocate the niche activities, and to tap into the processes of central government policy making and grant application. Without these informal and strategic relations, it would be much more difficult for niche actors to conduct in-depth negotiation with bureaucrats and complete an attractive grant proposal.

It is also true that local issues often stretch across the boundaries of government departments. The different jurisdictions of different areas create varying levels of criteria for grant application. Through the project proposal grant by the central government, the multi-niche scaled up to an upper governmental level. The contribution of this case for the discussion of scaling up for transition, in a practical sense, is that a strategic funding scheme can leverage the scaling up of niche-innovations by encouraging experimental integration which goes across different domains and multiple regimes. This result illustrates the role of multi-scalar interactions across multiple tiers of government (cf. Wittmayer and Loorbach 2016, Chap. 2, this volume).

As a whole, the case of the Higashiomi Welfare Mall, in hindsight, shares many process similarities with transition management, including the presence of strategic, tactical and operational activities. But most importantly, it underscores the relevance of *reflexive activities*, in the guise of the learning networks that not only experimented, but that also were able to discuss landscape changes and could strike new relations both across multiple domains and across multiple scale levels.

One bureaucrat from the Ministry of Environment, with no direct control over food, energy, or elderly care, also has expressed interest in the Welfare Mall. In a retrospective sense, the innovative projects were based on an informal "transition arena" and created by niche-level networks, not by formal and often sectionalised government policies, while at the same time being supported by the various funding schemes from across the Ministries. This enabled the achievement of the Welfare Mall, with an impact beyond the small local community, at different policy fields and Ministries. Officials in Ministry of Environment decided to subsidise planning of the second mall in another part of the city partially because they thought local self-reliance for food, energy, and elderly care may foster low-carbon development. The innovative challenge at the local niche level is now adopted by broader central government ministries as a strategy for countering landscape pressures.

References

Altieri MA (1985) Agroecology: the science of sustainable agriculture. Westview Press, Boulder

Coenen L, Benneworth P, Truffer B (2012) Toward a spatial perspective on sustainability transitions. Res Policy 41:968–979

Geels FW (2004) From sectoral systems of innovation to socio-technical systems: insights about dynamics and change from sociology and institutional theory. Res Policy 33(6-7):897–920

Geels FW (2007) Analyzing the breakthrough of Rock'n'roll (1930–1970): multi-regime interactions and reconfiguration in the multi-level perspective. Technol Forecast Soc Chang 74(8):$32#1411–1431

Geels FW (2011) The multi-level perspective on sustainability transitions: responses to seven criticisms. Environ Innov Societal Transit 1(1):24–40

Grin J, Rotmans J, Schot J, Geels i F, Loorbach D (2010) Transitions to sustainable development: part 1. New directions in the study of long term transformative change. Routledge, New York

Kemp R, Schot J, Hoogma R (1998) Regime shifts to sustainability through processes of niche formation: the approach of strategic niche management. Technol Anal Strat Manag 10(2):175–196

Loorbach D (2007) Transition management. New mode of governance for sustainable development. PhD thesis. Erasmus University Rotterdam

Loorbach D (2010) Transition management for sustainable development: a prescriptive, complexity-based governance framework. Governance 23(1):161–183

Loorbach D, Rotmans J (2006) Managing transitions for sustainable development. In: Olsthoorn X, Wieczorek AJ (eds) Understanding industrial transformation: views from different disciplines. Springer, Dordrecht

Lovely Town Eco Club Aito (2012) Mi-chi-bu-shi-n: town planning through strength, intellect, and people's power (in Japanese)

Ministry of Internal Affairs and Communications (2011) Discussion papers for green and local autonomy model cities, from April 28, 2010 to March 1, 2011 (in Japanese). main_sosiki/kenkyu/bunken_kaikaku/index.html

Mizuguchi S (2013) Higashiomi's Welfare Mall, challenge for local self-reliance for food, energy and care (in Japanese). August issue of Housing Tribune Magazine, Sohjusha Publishing, Japan

Naito M (2012) Vision and method of friendly to nature society, practice at Shiga: report to Socio-science Foundation of Japan (in Japanese)

Nill J, Kemp R (2009) Evolutionary approaches for sustainable innovation policies: from niche to paradigm? Res Policy 38(4):668–680

Nishimura T, Nomura M (2013) Citizen's power cooperative at welfare mall (in Japanese). Jpn Environ Technol Acad J 13:92–93

Nishimura T, Yamaguchi M (2011) Feasibility study on regenerating local forest for fuel through collaborative planning with citizens (in Japanese). Jpn Environ Technol Acad J 11:88–89

Raven R, Verbong G (2007) Multi-regime interactions in the Dutch energy sector: the case of combined heat and power technologies in the Netherlands 1970–2000. Technol Anal Strat Manag 19(4):491–507

Regeer B, Mager S, Van Oorsouw Y (2011) Licence to grow, innovating sustainable development by connecting values. VU University Press, Amsterdam

Rip A (2006) A co-evolutionary approach to reflexive governance – and its ironies. In: Voss J, Bauknecht D, Kemp R (eds) Reflexive governance for sustainable development. Edward Elgar, Cheltenham

Saito Y, Auestad KA, Waerness K (2010) Meeting the challenges of elderly care: Japan and Norway. Kyoto University Press, Kyoto

Schot J, Geels FW (2008) Strategic niche management and sustainable innovation journeys: theory, findings, research agenda, and policy. Technol Anal Strat Manag 20(5):537–554

Schot JW, Hoogma R, Elzen B (1994) Strategies for shifting technological systems: the case of the automobile system. Futures 26(10):1060–1076

Vernay AL, Pesch U, Van Bueren E et al (2014) Niche entrepreneurs in sustainable urban development. Paper presented at the 5th international sustainability transitions conference, Utrecht

Wittmayer JM, Loorbach D (2016) Governing transitions in cities: fostering alternative ideas, practices and actors, Chapter 2. In: Loorbach D, Wittmayer J, Shiroyama H, Fujino J, Mizuguchi S (eds) Governance of urban sustainability transitions, pp 17–36. Springer, Tokyo (this volume)

Chapter 6
Ghent: Fostering a Climate for Transition

Katharina Hölscher, Chris Roorda, and Frank Nevens

Abstract In 2011, the city administration of Ghent started up the 'climate arena' based on transition management (TM) to develop a plan and implement actions to work towards its ambition of climate neutrality by 2050 and to involve actors from the city in doing so. This chapter analyses the empowerment of the involved actors to discern long-term commitment to a transition to climate neutrality and draw lessons for TM. TM is found to yield initial 'seeds of change' by creating new contacts and synergies, broadening problem perceptions, and contributing to intrinsic motivation to pursue the transition towards climate neutrality. There is a need to differentiate between the arena participants and involved policy officers: TM appears as effective tool to open up policy officers for more open and co-creative approaches as well as cross-departmental collaboration by broadened problem foci. Challenges regarding the longevity of empowerment effects on the arena participants point to lessons for TM, including the creation of space in the city administration before the process and facilitation 'beyond the arena'.

Keywords Urban governance • Climate neutrality • Empowerment • Transition management • Co-creation • Change agents • Climate governance • Low-carbon cities

K. Hölscher (✉) • C. Roorda
DRIFT, Dutch Research Institute for Transitions, Erasmus University Rotterdam, Rotterdam, The Netherlands
e-mail: holscher@drift.eur.nl

F. Nevens
VITO, Flemish Institute for Technological Research, Mol, Belgium

D. Loorbach et al. (eds.), *Governance of Urban Sustainability Transitions*, Theory and Practice of Urban Sustainability Transitions,
DOI 10.1007/978-4-431-55426-4_6

6.1 Introduction

6.1.1 The Climate Neutrality Ambition of the City of Ghent

In 2007, the city council of Ghent had taken up the ambition to become climate neutral by 2050. The council simultaneously acknowledged that this ambition could only be achieved by involving the urban population because it did not find the city's policy and policy options sufficiently effective. To explore what a climate-neutral future would actually look like, to understand what is needed to achieve this ambition. and to involve actors from the city into developing and implementing actions, Ghent embraced transition management (TM) in 2010. A 'climate arena' was created that provided an informal but well-organised space to a group of 17 change agents from diverse backgrounds (Nevens and Roorda 2013). These arena participants were challenged to think and work beyond 'business as usual' in developing a vision and actions for a climate-neutral city as well as to take ownership of this process and actively pursue a climate-neutral Ghent. In this chapter, we explore the extent to which this TM intervention has created a common commitment of arena participants and city administration to a climate-neutral future of Ghent. We do so by assessing a TM 'claim', namely to influence societal processes of change towards sustainability by *empowering* societal actors "in such a way that people themselves shape sustainability in their own environments, and in doing so contribute to the desired transitions towards sustainability" (Loorbach 2007, p. 284). In other words, the involved change agents should become empowered through typical TM elements such as a protected space, interaction between niche-actors and regime-actors, and creating a systemic understanding of current challenges and future opportunities. It is often assumed that this approach sows some sort of 'seeds of change', also beyond the initial TM 'arena' setting; that a more empowered civil society develops, acting for pursuing a sustainability transition by engaging in actions and real-life transition experiments (Loorbach 2007, 2010; Rotmans and Loorbach 2011; Avelino and Rotmans 2009; Hölscher 2013; Hölscher et al. 2014).

The empowerment concept is intrinsically linked to the shift from government to governance in environmental decision making, of which private actor engagement is a main characteristic (Davies and White 2012; Driessen et al. 2012). Participation presumably provides opportunities for empowerment by creating room for learning and clarification of different views and interests; this might increase the acceptance of decisions, foster commitment to the implementation, and develop and expand capacities (Ansell and Gash 2008; Koontz and Thomas 2006; Newig and Fritsch 2009; Healey 2006). However, increasing emphasis on private actor engagement also raises criticism as to whether it solely transfers responsibility for action from public to private actors, or whether it risks dominance by powerful interests (Swyngedouw 2005; Lemos and Agrawal 2006; Fung and Wright 2001). These aspects entail potential *dis*empowerment and hence they can prove challenging in the TM context (Avelino 2009; 2011). This also refers to the Ghent case, in which the climate arena comprised diverse urban actors and in which the process was guided by a transition team that consisted of policy officers from the city administration.

6.1.2 A Focus on Empowerment by TM: The Research Questions

Acknowledging that 'empowerment' might still be a very ambiguous concept, transitions scholars recently started to theoretically and empirically explore how TM approaches deploy the concept (Avelino 2009; 2011; Hölscher et al. 2014). More insights are still direly needed to inform better understanding and derive lessons for future implementations. We draw on these earlier efforts to analyse the results of the climate arena in Ghent in terms of the empowerment of the involved change agents. In this analysis, we focus on two particularities. First, empowerment of change agents has both short-term and long-term connotations. Although the former relates more directly to the empowerment of the participants in the arena process, the latter refers to the long-term establishment of a 'climate for transition' and a somewhat self-propelling process by which the direction developed in the arena is pursued. In TM literature, there are thus far few insights into the longer-term dimension of TM processes 'beyond the arena.' Second, based on observations throughout and after the climate arena in Ghent, we aim to also account for the potential benefits of involving policy officials in the urban transition team. We hypothesise that they might, and even should, also experience empowerment to pursue the ambition of climate neutrality together with the arena participants.

Therefore, we project the following research questions on the Ghent case:

1. *To what extent did the actors involved in the TM process experience (an increased) short- and long-term empowerment for contributing to a transition towards a climate-neutral city?*
2. *Which lessons can be drawn for TM with regard to this empowerment of actors?*

6.1.3 Methodology

We undertake a qualitative in-depth case study (Gerring 2004) in this chapter. Our analysis is based on diverse data sources that cover different points in time of the TM process in Ghent: throughout the arena process (January 2011–January 2012) and 1.5 years after the process ended (May 2013). On-going unstructured observations are made about the follow-up of the process. This follow-up enables in-depth insights into the process implementation, results, and a comparison between short-term and long-term empowerment. In the course of desk research, we reviewed TM literature to derive the criteria for analysing the TM process and its results. For the analysis of the content itself, desk research covered official city documents, media sources such as Internet sites, press articles, brochures, and grey literature (e.g., project progress reports, internal research reports, minutes of arena meetings). Empirical research included several rounds of interviews and surveys by the authors of this chapter with the involved policy officials and arena participants. Semi-structured interviews with open questions were held before the arena process, throughout, immediately after the process, and 1.5 years after the process. Surveys were conducted immediately and 1.5 years after the process. Because of the involvement of two of the authors in the process implementation as external TM experts, additional data were collected through observations at arena and transition team meetings and personal communications. Partial research results are presented in two master's theses (Hölscher 2013; Maas et al. 2012) and scientific articles (Nevens et al. 2013; Nevens and Roorda 2013; Hölscher et al. 2014), as well as process reports (Roorda et al. 2014b; Roorda and Wittmayer 2014) and working papers (Roorda et al. 2014a; Wittmayer et al. 2011). The surveys were analysed by calculating both the average and mode to establish the most frequent value on the one hand and the general tendency on the other (Bryman 2004). To derive conclusions on empowerment levels, interviews were analysed based on previously established criteria and their operationalisation (Hölscher 2013).

6.1.4 Outline of This Chapter

In what follows, we first introduce the climate change governance context in Ghent (Sect. 6.2) before we describe the TM arena process and its outcomes (Sect. 6.3). In Sect. 6.4, we reflect upon the issue of empowerment of the involved actors. Section 6.5 holds conclusions and the lessons for TM, in particular with regard to a further 'beyond the arena' conceptualization.

6.2 The Climate Change Governance Context in Ghent

6.2.1 Historical and Political Context

Ghent is the second largest city in Flanders, the northern region of Belgium, and home to more than 247,000 inhabitants. The city's history reaches back to around 500 B.C., when Ghent was located in a river-connected marsh area, unsuitable for agriculture but fit for sheep breeding, which led to a flourishing wool industry. From around 1000 to 1500 A.D., it was one of the major industry and trade cities in Europe. After a period of decline, a new economic upswing was induced by industrialised linen and cotton production, making Ghent one of the first cities on the European mainland to enter the Industrial Revolution. Around that time, it also evolved into a prominent university city and its port opened up, becoming the second largest Belgian harbor. In the 1970s, having developed into an industrious, yet grey and polluted city, Ghent underwent another strong economic shift with an increasing share of service- and knowledge-based activities. Local politicians took important decisions to improve the situation in the city, by, for instance, demarcating part of the center as a traffic-free zone and improving the quality of the surface waters. Furthermore, the city started cultural activities, most famously the Ghent Festival ('Gentse Feesten'), which gradually evolved into a yearly city-capturing summer event that attracts 1.7 million visitors per year. Specific elements of the past also characterise the city today, such as the historical center, a large and lively student community, rich cultural life, and significant harbour activities (Maas 2011). Politically, the city resembles an island in Flanders: in many municipalities a right-wing tendency characterised the 2012 elections, whereas in Ghent the Social-Democrats and the Green Party emerged as winners. There is a general image of the city being progressive and dynamic.

6.2.2 Urban Climate Governance in Ghent

In 1996, Ghent had already joined the Climate Alliance, a network of European cities and municipalities to reduce greenhouse emissions. From 1998 onwards the city regularly drafted environmental and energy policy plans. In 2007, instigated by the alderman for environment, urban development, and dwellings, the city council expressed the strong ambition to become climate neutral by 2050 and, for that cause, to reduce CO_2 emissions by 50 % by 2030 (Balthazar 2008). Two ambitious policy plans regarding energy and climate were established in 2008. Ghent's mayor signed the European Covenant of Mayors in 2009, thereby declaring its intent to curb its CO_2 emissions beyond the 20 % reduction target set by the EU for 2020 and beyond the targets of the national and Flemish governments (EC 2012; Maas 2011). Besides a general orientation of the environmental department towards climate challenges, a climate team of eight administrators was set up. The strategy of this team was twofold: 'to study and to involve.' Hence, the climate team initiated quantitative studies (e.g., a baseline inventory of greenhouse gas emissions and an abatement cost curve for energy-saving options) and endeavoured to publish sensitising information in appealing ways (e.g., an online aerial heat loss photo enabling citizens to observe the performance of their own houses).

From the start, the city administration acknowledged that the expressed climate ambition could only be achieved by decisive action from all actors in the city. Therefore, the Ghent Climate Alliance ('Gents Klimaatverbond') was set up in 2009. Under that umbrella, the city attempted to involve citizens and companies in drafting a new, ambitious climate plan that would sketch the development towards a climate-neutral city. Additionally, in 2010, the environmental department tendered an assignment to facilitate that involvement of actors. The approach of the winning consortium was based on the principles of TM and advocated more visionary thinking and creative idea generation for experiments. This process coincided with and was later linked to the Interreg IVb-funded MUSIC project (Mitigation in Urban Areas: Solutions for Innovative cities; see Wittmayer et al. (Chap. 3, this volume) (in which Ghent was one of the partner cities), resulting in the set-up of the 'climate arena'.

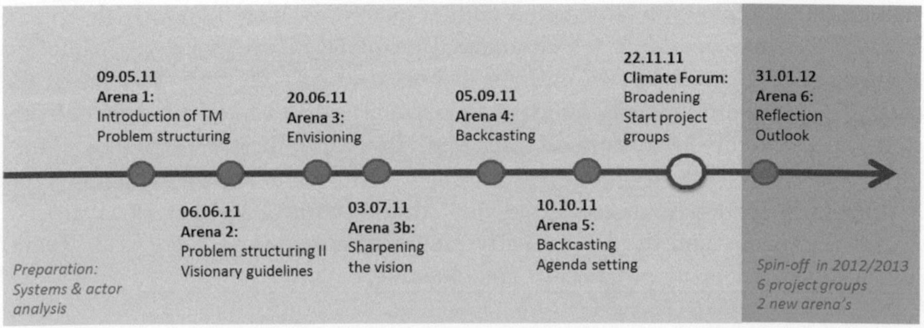

Fig. 6.1 Timeline of the climate arena process in Ghent (From Roorda and Wittmayer 2014)

6.3 The Climate Arena Process in Ghent

In the course of the MUSIC project, TM was operationalised for the urban context (Nevens et al. 2013; Roorda et al. 2014b). Figure 6.1 provides an overview of the arena process in Ghent; the following sections describe the preparation, implementation, outcomes, and follow-up of the different steps in the process.

6.3.1 *Preparation*

6.3.1.1 Transition Team

After some initial restructuring, the transition team consisted of two policy officers from the city's environmental department, two members[1] of the facilitating consortium, and one extra TM expert.[2] The role of the transition team in the process is to prepare and guide the arena meetings and select the participants, among other tasks. In spite of the time pressure of their 'business as usual duties,' the two motivated transition team policy officers could dedicate significant time to the climate arena process (on average about 1 day per week). City administration colleagues did not provide much support, partly because of skepticism and doubts about the process and its relevancy for their work.

The need for flexibility in the arena process and adapting it to the relatively open and unpredictable dynamics was, but limitedly, allowed for by the director of the environmental department and the responsible alderman. They were supportive of the process, yet the tradition of a controlling attitude and the political commitment put much time pressure on the process from the start, which significantly limited the co-creation of the process by arena participants.

[1] Including the third author of this chapter.

[2] The second author of this chapter.

6.3.1.2 Systems Analysis

In the first months of the process, the transition team—with profound support by a master's student (Maas 2011; Maas et al. 2012)—conducted a systems analysis that included an assessment of the social, ecological, and economic aspects of Ghent by means of the SCENE model (Grosskurth and Rotmans 2005), a historical sketch and a trend analysis. The analysis created a useful overview of the system, identifying problem areas and city trumps that could be used for future scenarios, as well as highlighting interlinkages between environmental, economic, and social themes and hence the cross-departmental, wicked, and very systemic nature of the climate issue (Maas 2011).

6.3.1.3 Formation of the Arena Group

Concomitantly, the transition team performed an actor analysis to identify whom to interview to feed the systems analysis and to select participants for the arena: starting from a pragmatic long list (acknowledging that a vast potential of change agents in the city was yet unknown to the administration), coming, through actor mapping, to an informed 'short list' of actors, and ending with 20 invitees. The result was a final core arena group of 17 committed members with wide variety in age, gender, and backgrounds, yet all concerned with sustainability.

6.3.2 Climate Arena Meetings

6.3.2.1 Problem Structuring

The insights from the systems analysis were presented and discussed during the first arena meeting. Seven introduced 'trumps' of the city (authenticity, labor, water, knowledge, livability, sympathetic, energetic) (Nevens and Roorda 2013) inspired a discussion in which the initial climate problem perspectives opened up with issues such as energy poverty and general well-being of citizens (Maas 2011; Maas et al. 2012), hence revealing interesting and surprising connections and potential win–wins.

6.3.2.2 Envisioning

In the second and third arena meetings, the arena members defined basic principles for a sustainable Ghent in 2050 (e.g., a car-free inner city, independency of fossil fuels, companies delivering societal return to the city and its citizens). This goal allowed for both 'abstract' future thinking and the emergence of new ideas for the short term. Although too little time might have been devoted to result in an elaborated vision, some participants also urged not spending too much time on this stage as they felt this step was rather abstract and too free of engagement.

6.3.2.3 Backcasting and Agenda Setting

The fourth arena meeting centered on participatory backcasting. Four subgroups identified 20 transition pathways[3] towards the envisioned future and determined milestones and actions for a few of these pathways. The transition team policy officers complemented the remaining pathways. New actors were invited to the fifth arena meeting, aimed at setting priorities between the pathways, because of their link to the transition pathway themes. Five participants took a leadership role for a specific pathway, willing to further chair a subgroup to formulate the real-life transition experiments that they would like to realise.

By the end of the fifth meeting, a transition team policy officer announced that this had been the final meeting. This closure came as an unpleasant surprise to some participants, as they felt the need for more time and effort to continue what had been started. The extensive personal contact of the transition team with the arena participants on an individual basis resolved some of the discomfort and secured further commitment, at least for contributing to the subsequent event at which the ideas were presented, the Climate Forum.

6.3.3 Outcomes of the Arena Process

According to TM literature, a transition arena should eventually result in actions that target the envisioned sustainability transition by challenging discourses and introducing new practises (Loorbach 2007, 2010; Frantzeskaki et al. 2012). This growth fundamentally depends upon the dissemination of the arena ideas and the engagement of new actors as well as the development and implementation of initiatives that actualize them. Although the empowerment of the involved actors is closely linked to these activities and can also be considered an outcome of the arena process, it is analysed in depth in Sect. 6.4.

6.3.3.1 Dissemination and Engagement

In late 2011 and early 2012, the environmental department and the arena participants disseminated their ideas and engaged a broader group of actors. The major

[3] The pathways are divided into four clusters: 'Creating added value locally with an opening to the world,' 'Ghent, good to live in,' 'Energetic city: intelligent cycles,' and 'Ghent's citizens feel at home in their city.'

event organized for this purpose was the Climate Forum, for which the arena participants had to invite (‘headhunt’) people who could contribute to the emerging transition experiments. More than 100 highly motivated people attended this event. They were informed about the arena results thus far and were involved in Climate Working Groups (CWGs) that formed around particular themes and further elaborated or developed additional transition experiments.

By the end of 2011, the ideas from the arena were published in the form of ‘transition magazines’ to promote the vision of the arena and lobby for support. For each arena participant a personalized edition was made, which included a photo of the participant on the cover and a personal interview next to an introduction of the process and the vision.

To target politicians, the arena work and results were presented to the city council by a small group of arena participants and by the transition team. Three political parties for the council elections in late 2012 showed a high ambition level regarding climate measures, and their election programs included specific ideas from the arena. It is not possible to assess the exact influence of the arena (in general, a high level of political commitment to the climate neutrality ambition existed); yet by visibly entering the political forum, it potentially reinforced a further elaboration of policy measures and a significantly higher budget for energy and climate policies.

6.3.3.2 Development and Implementation of Activities

Most of the (spinoff) activities (Table 6.1) were elaborated and implemented in the CWGs. These were initiated at the Climate Forum and consisted of both arena participants and gradually added new actors. Next to the CWGs, two initiatives (Mobility Arena and Ghent University transition) were inspired by the TM methodology itself and started up similar arena processes in a different context. Additionally, the Green Track initiative (climate ambitions in the cultural sector) started up and was linked to the process. The city administration took a supportive, but limited role: thinking along, creating connections, and offering logistic support. Most initiatives needed time to mature before they could actually be deployed; one did not succeed the attempt. Half of them did not directly relate to the operational level, and thus did not represent ‘classical’ transition experiments as conceptualized in the transition literature (Van den Bosch 2010), but rather aimed at further engagement and elaboration of more specific topics. In that, the ‘spinoff’ activities of the arena enabled more actors to see their role in working towards a low-carbon future, to explore new practices and paradigms, and to reach a broader audience. From the broadening of the climate neutrality theme, the activities take very diverse entry points to approach the issue, although their direct contribution to climate neutrality is hardly measurable.

Table 6.1 Overview of spinoff initiatives of the climate arena process

Initiative	Relationship to arena process	Outreach
Carrot mob	CWG 'consumer pushes the market'	A 1-day carrot mob attracted 938 mobbers and made the targeted supermarket invest €10,000 in sustainability measures
Urban farming project with children	Arena participant active in CWG 'urban agriculture'	In collaboration with social youth organization; project with a small group of children, involving growing and harvesting mushrooms, and having dinner with the children's families
Valorization of sewage water projects	CWG 'valorization of sewage water'	Development and implementation of a business case for city district 'Old Docs' (350 dwellings) on the transport of organic waste through the sewage system and the use of sewage water to produce heat, biogas, nutrients, and water
Energy efficiency for businesses	CWG 'energy efficiency for businesses'	Financed by the city administration; aimed at stimulating and guiding small and medium-sized enterprises (SMEs) to structurally work on energy efficiency; the project started with a group of 10 SMEs
Blue economy	CWG 'Blue economy'	The ambition was to find a company that would set up an iconic project in line with the Blue economy principles to enhance ecosystems. The CWG was unsuccessful in finding a focus and partners.
Green Track	Five art organizations created CWG 'Green Track' in linkage to climate arena	Aim is to monitor CO_2 emissions of the organizations that are participating; 30 cultural organizations joined. Brussels, Leuven, and Antwerp also participate in this project now.
Mobility arena	Mobility and environmental departments of the administration joined forces; involvement of some arena participants	New arena process to explore the future of mobility in Ghent; launch of a very well received vision and agenda 'The bike of Troy'; three iconic projects as spin-off, most notably the 'Leefstraat' (two streets became car free for 1 month in 2013, more streets joined in 2014); regular follow-up meetings in 'mobility cafes'
Transition UGent	Arena participants initiated and implemented process	New arena process initiated by the University of Ghent to explore how the university could become sustainable; 120 students, professors, and administrative staff members participated; final plan 'Transition UGent–together for a sustainable university' formally approved by the board of the university; spinoff activities are now starting up (e.g., garden in the faculty of engineers); follow-up sustainability platform

Adapted from Hölscher (2013)

6.4 Empowerment of Change Agents?

Hölscher et al. (2014) define "an empowerment process as designed according to principles that should develop or enhance political, economic, cognitive, social and psychological capacities of individuals to achieve desired ends, resulting in such increased capacities (empowerment outcome)." This definition links the empowering process to the level of empowered outcome, highlighting the importance of analyzing both in relation to each other. In TM, the arena process should empower the participants to pursue the envisioned sustainability transition by engaging new actors and by developing and implementing activities and experiments. Thus, empowerment also relates to the long-term pursuit of a sustainability transition to achieve the ambition of climate neutrality in the case of Ghent. Three dimensions of 'empowered' outcome are distinguished: *political*, *cognitive*, and *psychological*. In the following analysis of these dimensions, we differentiate between short-term (immediately after the arena process) and longer-term (1.5 years after the arena process) empowerment. Additionally, we distinguish between arena participants and policy officers because of their different roles in the process. Importantly, we hereby do not intend to imply that any actor was a change agent at the start of the process, using 'change agent' as essentialist criteria, but rather apply the empowerment lens to explore whether the process contributed to them acting as change agents.[4]

6.4.1 Political Empowerment: Network Ties and Synergies

A specific transition arena objective is to connect a heterogeneous group of actors and hence induce innovative networks ties and synergies that can initiate and accelerate the envisioned transition (Loorbach 2007, 2010; Nevens and Roorda 2013). Such cooperation for mutual benefits and pooling of resources (e.g., economic, time, technical and logistical support, expertise) can be denoted as political empowerment (Hölscher et al. 2014). Indeed, the arena participants and policy officers of the climate arena in Ghent considered the new relationships, established in the open and amiable arena atmosphere, amongst the most important results (Table 6.2). With the later broadening activities this network dynamic proved transmittable and expanded towards newly engaged actors. Although in due course

[4] This differentiation is crucial because TM usually intends to involve change agents in the arena and empowerment therefore targets change agents. This is in line with the definition of change agents in the urban context as "urban frontrunners" (Wittmayer et al. 2011, p. 3) who understand aspects of urban wicked problems and have ideas on long-term images of the city, are innovative thinkers or activists in urban development, are network brokers, and have a genuine and passionate interest in the society and commitment to its sustainability.

Table 6.2 Short- and long-term political empowerment of actors in Ghent

	Arena participants	Policy officers
Short-term political empowerment	Close relationships have been established among the participants and towards the involved policy officers, partially even "friends were made".	The involved policy officers could use the process to build up new network ties not only to the arena participants and newly engaged local actors, but also to policy officers from other departments and to TM experts and other practitioners.
Long-term political empowerment	The established network ties between participants, towards policy officers and newly engaged actors remained largely very loose as no initiative for follow-up meetings was taken. Nevertheless, several participants were able to use these ties, and especially the ties to the policy officers from the administration for the initiation of joint activities. They are considered valuable for further knowledge exchange, when opportunities occur.	Most of the established network ties were maintained. They have helped to create greater momentum and leverage for the joint pursuit of the climate neutrality ambition of Ghent.

the links between the arena group members have weakened and remained rather loose, the ties towards and between the involved policy officers seem to have enabled synergies, stronger than before, for the further pursuit of a climate-neutral Ghent.

From the interviews and survey 1.5 years after the process and the ongoing observations we can discern that, in spite of their closeness during the arena meetings, the process did not generate lasting strong relationships among the arena participants. Nevertheless, the informal network of the arena was continually valued, especially with regards to those contacts relating to 'sectors' with which participants did not have prior relations. The lack of more durable links was attributed to the city administration, which was considered "the glue of the process" and "the reason you come together." Accordingly, the one-on-one contacts between participants and the city administration crucially contributed to the success of the spinoff initiatives, having served as a setting for consultation support, funding, and facilitation of the development and implementation of activities such as the carrot mob and the transition Ghent University process. Indeed, the connections that did last are related to the CWGs, which were typically connected to the professional backgrounds of their participants.

So far as the policy officers are concerned, the arena process has had a lasting impact: the involved officers created many new bonds with actors from the city. Additionally, cross-departmental considerations of climate neutrality and new bonds sought by the involved officers between (policy officers in) the various departments emerged and persisted. This change was most apparent in the mobility arena, which was jointly organized by the departments of mobility and environment, and in an envisioning process of the spatial planning department. It sparked an evident shift from thinking in terms of 'own territory/turf' to shared challenges and ambitions.

6.4.2 *Cognitive Empowerment: Understanding of Complexity and Behavioral Options*

TM pursues cognitive empowerment for actors to (better) understand and handle complexity and uncertainty as main characteristics of system(ic) change, to take up new perspectives, understand challenges, and to define specific roles in the broad action field (Loorbach 2007, 2010). This step is envisaged to be aided throughout the arena process, including the systems analysis, envisioning, and backcasting, the discussions in the heterogeneous frontrunner group, and the provision of additional knowledge to arena members (Hölscher et al. 2014).

In Ghent, arena participants and policy officers experienced the climate arena process as a significant cognitive stimulation because of the varied inputs, the lively and interesting discussion topics, and the introduction of ideas, topics, and perspectives that were new to them (Table 6.3).

Table 6.3 Short- and long-term cognitive empowerment of actors in Ghent

	Arena participants	Policy officers
Short-term cognitive empowerment	The participants obtained a greater understanding of the complexity of the climate issue and sustainability challenges and of the different perspectives on them. Additionally, they attained technical insights. Two participants stated to have acquired and implemented TM in their professional environments. Limited was the identification of own roles in the pursuit of climate neutrality.	The involved policy officers gained insights about participatory processes in general and the TM methodology specifically. In relation to this, the policy officers have learned from the arena participants, which increased their understanding of citizens' perspectives. The broadening of the climate issue opened up new ways of working with other departments.
Long-term cognitive empowerment	It is difficult to establish the long-term effects of cognitive empowerment. The participants continue to find the insights gained useful in their everyday life, yet their instrumental value seems to remain limited. In general, the fact that some CWGs matured into implementing activities only after some time points to longer-term effects of the process.	The policy officers applied and expanded their experience with TM throughout spinoff arena processes; this has opened up their way of working and is even spreading throughout the environmental department and the city administration as a whole, albeit still remaining in the margins. Cross-departmental linkages that can be exploited were expanded in pursuing a climate-neutral Ghent holistically.

Induced by the systems analysis and the open discussions, all participants and involved policy officers gained a broadened view on climate neutrality. They found various topics that were discussed in the arena group enriching: the theme of the blue economy, the importance of water cycles, the potential of city greening, etc. Most participants were surprised about the social dimension of climate concerns, which they had not considered much before. One participant summarized this experience as follows: "[...] for me initially some of the reactions of the people were difficult to understand. And during the process you grow to each other and you are able to look from their perspective."

Besides acquiring a broadened view on climate neutrality, the arena participants found that they have gained (technical) knowledge beyond their fields of usual expertise. Moreover, like the policy officers, some participants referred to the TM approach as yielding valuable knowledge and stimulating the implementation of (parts of) that knowledge in their further work. Especially for the policy officers involved, the climate arena represented a profound learning experience about co-creative approaches and stimulated a redefinition of their functions in relation to other actors in the city—a learning effect they try to bring into the city administration as a whole.

Several participants described a more general use of the broadened view they had acquired. According to one: "It helps me to motivate other people to have a more general view and to show people the limits of certain pathways." For the most part, however, the knowledge gained has been of intrinsic value rather than instrumental, meaning that the contribution to defining own roles and behavioral options in alignment with the envisioned transition was not prime. If behavioral options were induced they were generally greater for those participants who could make direct links to their professional backgrounds; others felt restrained by a lack of (mostly financial) resources. Still, the cognitive empowerment is noteworthy, as illustrated by a policy officer's statement that the arena can be framed as a method for participation where the level of knowledge of the participants is increased and complexity of the subject is acknowledged. According to her, other participation trajectories often fail because citizens do not have sufficient knowledge or only defend their own interests. The arena participants even repeatedly voiced the desire to receive more input from experts on the themes they have identified as relevant to climate neutrality, to develop better images of the future and actions for the present.

6.4.3 Psychological Empowerment: Intrinsic Motivation

A main ambition of the TM process is to intrinsically motivate actors to actively (as 'change agents') pursue the transition that has been jointly envisioned in the arena (Avelino 2011; Loorbach 2007, 2010; Rotmans and Loorbach 2011). Intrinsic motivation of actors to self-efficacy essentially depends on "interpretive processes through which individuals assess and evaluate their tasks, activities, and roles"

(Avelino 2009, p. 377; see also Thomas and Velthouse 1990). It is achieved by four dimensions of positive task assessment:

- Meaning (value attributed to a task)
- Competence (belief in capacities to fulfill a task)
- Impact (conviction to achieve intended effect of a task)
- Choice (sense of autonomy and self-determination in initiating and completing a task).

It is striking that although the participants were initially very motivated to continue the process, 1.5 years after the rather abrupt end of the arena meetings, this level of motivation has dropped considerably (Table 6.4). Participants without external support or options for professional linkages have often not pursued any further action. In contrast, the involved policy officers have attained and sustained a high level of motivation for achieving a climate-neutral Ghent and, additionally, for continuing participatory efforts–also in form of other TM processes. This situation has led to a continuation of follow-up activities, which are often led or facilitated by the city administration.

All participants had been highly committed to sustainability issues upon joining the climate arena. This attitude was initially fostered throughout the arena process and developed into a joint commitment and the implementation of several activities (Table 6.1). Most participants recounted in interviews and conversations

Table 6.4 Short- and long-term psychological empowerment of actors in Ghent

	Arena participants	Policy officers
Short-term psychological empowerment	Most participants were motivated and inspired to pursue the envisioned transition towards a climate-neutral Ghent by developing and implementing activities. The feeling of being in a group of like-minded propelled their sense of being able to do so.	The involved policy officers have – from the beginning – greatly supported the goal of climate neutrality for the city of Ghent. Their commitment to the follow-up of the arena process was at first limited. They took a secondary role, also because of the same capacity constraints that urged a withdrawal from the process.
Long-term psychological empowerment	In light of (perceived) lack of support by the city administration and capacity constraints, many participants started to doubt their competence and impact for significantly contributing to a climate-neutral Ghent. Most of them hold the city administration in charge for facilitating the on-going process and implementing measures. Mostly participants with professional linkages to the topics remained empowered.	The involved and additional policy officers continue to lobby and work for both TM as promising participatory methodology and a climate neutral city. However, they also see limitations of the process and doubt as to whether the administration can act as main facilitator, because of tensions with other roles of the municipality, capacity constraints, and lack of specific skills.

how envisioning a climate-neutral Ghent in 2050 with a group of like-minded participants was contagious, inspiring, and motivating. A participant stated: “Because then you were part of the process thinking, making the ideas and coming to a story and it’s part of your story. Even if it’s a small part, some of the ideas were yours and that’s why the end result is something that you carry with you.” In the long-term perspective it turned out that the meaning attributed to climate neutrality as the collective goal became limited. In this regard, it appears as a main challenge to sustain a sense of competence, impact, and choice. Statements such as “If the city administration drops out, there is no project anymore. [...] Because it is not a process that leads its own life. I don’t really believe in that. They should be the initiators and the motivators” closely linked to this. Participants thought that for a significant impact, far-reaching (political) decisions and support were necessary. Interestingly, especially for those participants working professionally in the sustainability field, the sense of competence and impact to contribute to a climate-neutral city was already quite high and remained that way. It has been largely those participants who have been able to initiate follow-up activities because they could link them to their professional background.

For the involved policy officers the broadening of the ambition has given greater meaning to the climate neutrality ambition, resulting from a more holistic approach that creates greater impact by cross-departmental spread and intense interaction and co-creation with stakeholders. These interactions were also seen as instruments to enrich and broaden the necessary competences in a way that potentially yields more impact; an insight shared by colleagues who “saw that it was a leverage in terms of budget cuts and less people, of being able to make things happen”. However, doubts persist as to whether this approach is applicable to each ‘domain’ (e.g., refurbishment of buildings).

6.5 Conclusions

6.5.1 Empowerment Through TM

The analysis of the TM process in Ghent offers elements that support the conclusion of TM positively contributing to the empowerment of participants with regard to the three distinguished dimensions. Hence, it potentially contributed to the realization of the envisaged climate neutrality ambitions and challenges. The climate arena has stimulated the participants and involved policy officers to start a learning trajectory about what climate neutrality means for the city and for them. It encouraged systemic thinking, enabling a diverse group to get a grip on the abstract, often technically framed topic by putting it in a broader societal context. Experiencing the process together has connected diverse actors more closely, creating a feeling of like-mindedness. The result is not-so-tangible but invaluable outcomes: a broadened perception and ownership of underlying problems and potential solutions as well as a loose network to which both participants and policy officers can revert if needed.

The empowerment impacts were for most arena participants not lasting, especially with regard to psychological empowerment: this might have several reasons. A main reason might have been the rushed and somewhat top-down process implementation, which did not allow for ownership to develop on the parts of the participants and limited the identification of the own roles and responsibilities in the pursuit of climate neutrality. It also hints at the limitations of a 'narrow' TM approach, centered on the transition arena and not directly looking for providing resources, for example. A more opened-up approach could tackle the obvious need for both the creation of space within the administration at the outset of the process and for continued efforts 'beyond the arena' to establish a durable empowerment momentum. On the other hand, for the involved policy officers from the city administration, the effect of empowerment was more persistent. The broadened view on climate neutrality, extending the view from 'own domain' to 'shared domain,' and the TM realm itself were taken along and spread towards other departments and initiatives. The policy officers involved act as change agents within their administration, lobbying for (the philosophy and attitudes of) TM and more co-creative approaches in general, for a broadened view on the climate issue and other matters.

All in all, the climate arena resulted in, albeit almost intangible, 'seeds of change' towards climate neutrality: a broader variety of actors having a part in the transition (e.g., through the spin-off activities), and insights from the problem framing and envisioning are taken up by other parties. In this way, the experience in Ghent shows that this arena process can give an (extra) impulse to a joint movement for a sustainable city. Genuine empowerment of actors to become effective change agents can be initiated/stimulated by TM. It appears, however, that the process implementation and some sort of follow-up 'beyond the arena' are crucially linked with a lasting success of TM.

6.5.2 Lessons for TM

TM is a framework with supporting tools, not a script or ready-made policy instrument that can be rolled out. Therefore, this instance of applying TM is unique, strongly influenced by the context and the people involved. Still, from the experiences in Ghent we can draw some general lessons about the application of the approach, in particular regarding the interrelation between process implementation, empowered outcomes, and follow-up action towards a sustainability transition.

Sound knowledge input and a thorough preparation of the content of the arena meetings appeared of great importance for cognitive and also psychological empowerment. The example of Ghent shows that the systems analysis and envisioning exercise stirred up the initially positive empowerment effects, serving as instruments for shared understanding, broadening perspectives, dialogue, and co-creation about a sustainable and climate-neutral future for the city.

The speed of process implementation proves a delicate issue, particularly with regard to psychological and political empowerment. The process does not only ask for investing time, but also for patience: ideas need time to evolve, new contacts need time to develop trust, conditions need time to change, and also opportunities are not always readily available. A rash process implementation might limit the participants' ownership over the process, especially when they are not involved in deciding on it, and might threaten the depth of the discussions and outputs. By contrast, the process needs to maintain a certain speed to sustain enthusiasm, create momentum, and seize opportunities. When new actors become involved at a later stage, specific project ideas from the transition agenda will evolve, as these actors will bring in their personal agendas and ambitions.

The case of Ghent also illustrates the challenging balance between 'steering'/'coordinating' and 'following'/'letting go'. The transition team led the structuring of the process and elaboration of the content, while verifying that very content continuously with the arena group. Although the participants appreciated this, too little space was left for their input. This restriction threatens to undermine ownership and the extent the participants identify their roles and responsibilities in the process, exacerbating long-term empowerment. At the same time, a certain degree of steering seems to be necessary for a level of sufficient motivation, collaboration, and knowledge. Leaving the steering component out too early entails the risk of losing the gradually increasing foundation that underpins empowerment for pursuing the envisaged sustainability transition, as also pertains to the follow-up of the process 'beyond the arena.' A minimum of facilitated follow-up seems quintessential, at least to provide a platform for synergy and reflection; this links to the levels of empowerment, as it would provide some sort of external motivation. A related lesson is that the involvement of policy officers in the transition team requires the creation of space before the arena process to acquaint them with the co-creative mindset.

Regarding political empowerment, it should be acknowledged that the activities after and next to the arena process ask for different capacities of the involved actors than in the arena process itself. We can now see 'transition networks' emerging around the CWGs. Other networks and synergies are sought: although the arena targeted people on a personal level and focused on change agents with the ability to think beyond 'business as usual,' the transition networks mainly involve actors who can link to a professional level and have the capacity of translating new ideas to new practices or 'selling' ideas to a broader audience. In Ghent, this shift was already anticipated by inviting new actors, including colleagues from other departments, to the later arena meetings and especially to the Climate Forum.

6.5.3 Discussion and Outlook

In this chapter, we focussed on how the TM approach empowers actors involved during and after an arena process. There are some clear limitations to the research approach taken. The 'long term' we assess is only 1.5 years, which still seems exceedingly short compared to the aim of a long-term transition that takes one to two generations. Moreover, the analysis of empowerment effects in based on the self-assessment by the actors involved. More crucial appears the limitation of the analysis to actors involved in the arena process, not to those who became involved later (i.e., in the CWGs) or were influenced/inspired by the outputs of the process (i.e., policy officers from other departments, politicians). These actors could more easily connect their personal ambitions and professional agendas to the ideas taken forward. Despite these limitations, the elaborate data collection allows for drawing some clear insights on TM with regard to the empowerment of actors to contribute to sustainability transitions.

Our analysis opens up several (research) issues for the practice of TM. The empowerment of the policy officers involved and the associated developments inspired within the city administration requires further attention within the TM research and practice. A key question is how to further transform (municipal) organizations to create space for going beyond 'business as usual' while they typically work on confirming the status quo. TM has shown to be an instrument for creating such space, although the policy officers involved struggled with re-inventing their own roles and with strong criticism from their colleagues, as TM does not suggest the usual degree of control over the process and its outcomes and is thus at odds with the regular policy logic. Such efforts could build on organizational learning literature and inquire how learning takes place within the administration and the effects on its relationships to actors from the city. A related issue that appeared from the analysis regards the necessary balancing acts that have been carved out as being inherent to TM. Future research should focus on in how far trade-offs can be avoided or how these could be handled.

An additional point for research is the conceptualization of the follow-up of the arena. Our analysis indicates that the determination of a strategy especially for the getting-into-action part is crucial to increase the effectiveness of the arena; this concerns the long-term empowerment of involved actors and how empowerment could be sustained, and the relationship between city administration and arena participants, as well as successively involved actors. Currently, settings for empowering processes that enable social learning for accelerating transitions are the transition team, transition arena, and transition experiments. The analysis implies that more flexible forms of transition events are required to create space for new paradigms and practices. Similarly, it refers to questions of resources and power: how it relates to the arena participants and their resources and professional backgrounds, the interactions between arena participants and city administration, and the power dynamics throughout sustainability transitions.

Acknowledgements This research has been conducted as part of the EU-funded projects MUSIC (www.themusicproject.eu) and IMPRESSIONS (www.impressions-project.eu). The authors thank the editors of this book and the reviewers who have provided valuable feedback to this chapter.

References

Ansell C, Gash A (2008) Collaborative governance in theory and practice. JPART 18:543–571. doi:10.1093/jopart/mum032

Avelino F (2009) Empowerment and the challenge of applying transition management to ongoing projects. Policy Sci 42(4):369–390. doi:10.1007/s11077-009-9102-6

Avelino F (2011) Power in transition. Empowering discourses on sustainability transitions. Dissertation, Erasmus University Rotterdam

Avelino F, Rotmans J (2009) A dynamic conceptualisation of power for sustainability research. J Clean Prod 19(8):796–804. doi:10.1177/1368431009349830

Balthazar T (2008) Lokaal Klimaatplan Gent 2008–2020. 105 acties voor een klimaatneutrale stad. http://www.gent.be/docs/Departement%20Milieu,%20Groen%20en%20Gezondheid/Milieudienst/20080605%20Lokaal%20klimaatplan%20Gent%202008-2020%20105%20actiepunten.pdf. Accessed 18 Sept 2014

Bryman A (2004) Social research methods. Oxford University Press, Oxford

Davies AL, White RM (2012) Collaboration in natural resource governance: reconciling stakeholder expectations in deer management in Scotland. J Environ Manag 112:160–169. doi:10.1016/j.jenvman.2012.07.032

Driessen PPJ, Dieperink C, van Laerhoven F, Runhaar HAC, Vermeulen WJV (2012) Towards a conceptual framework for the study of shifts in modes of environmental governance. Experiences from the Netherlands. Environ Policy Gov 22(3):143–160. doi:10.1002/eet.1580

EC European Commission (2012) What is the EU doing about climate change? http://ec.europa.eu/clima/policies/brief/eu/index_en.htm. Accessed 6 Sept 2014

Frantzeskaki N, Loorbach D, Meadowcroft J (2012) Governing societal transitions to sustainability. Int J Sustain Dev 15(1/2):19–36

Fung A, Wright EO (2001) Deepening democracy: innovations in empowered local governance. Politics Soc 29(1):5–41

Gerring J (2004) What is a case study and what is it good for? Am Polit Sci Rev 98(2):341–354

Grosskurth J, Rotmans J (2005) The SCENE model: getting a grip on sustainable development in policy making. Environ Dev Sustain 7(1):133–149

Healey P (2006) Collaborative planning. Shaping places in fragmented societies. Macmillan, London

Hölscher K (2013) Seeds of change? Evaluating the transition arena for urban climate governance in Ghent and Ludwigsburg. Master's thesis, Utrecht University

Hölscher K, Wittmayer J, Avelino F, Giezen M (2014) Exploring (dis-) empowerment in transition management. The case of urban transition management in the MUSIC-project. Paper presented at NWO-JSPS seminar studying sustainability transitions in welfare states, Rotterdam, 10–13 Sept 2014

Koontz TM, Thomas CW (2006) What do we know and need to know about the environmental outcomes of collaborative management? Public Admin Rev 66:111–121. doi:10.1111/j.1540-6210.2006.00671.x

Lemos MC, Agrawal A (2006) Environmental governance. Annu Rev Environ Resour 31:31297–31325. doi:10.1146/annurev.energy.31.042605.135621

Loorbach D (2007) Transition management. New mode of governance for sustainable development. Dissertation, Erasmus University Rotterdam

Loorbach D (2010) Transition management for sustainable development: a prescriptive, complexity-based governance framework. Governance 23(1):161–183

Maas S (2011) Making sense of complex systems. A systems analysis methodology for Urban transition management: the case of Ghent and its climate ambitions. Master's thesis, Wageningen University

Maas S, Fortuin K, Frantzeskaki N et al (2012) A systems analysis methodology for exploring urban sustainability transitions. Exploring challenges and opportunities for urban sustainability transitions in Ghent and Aberdeen. Paper presented at 3rd international conference of sustainability transitions, Copenhagen, 26–28 Aug 2012

Nevens F, Roorda C (2013) A climate of change: a transition approach for carbon neutrality in the city of Ghent (Belgium). Sustain Cities Soc 10:112–121. doi:10.1016/j.scs.2013.06.001, 2014

Nevens F, Frantzeskaki N, Loorbach D, Gorissen L (2013) Urban transition labs: co-creating transformative action for sustainable cities. J Clean Prod 50:111–122. doi:10.1016/j.jclepro.2012.12.001

Newig J, Fritsch O (2009) Environmental governance: participatory, multi-level – and effective? Environ Policy Gov 19:197–214. doi:10.1002/eet.509

Roorda C, Wittmayer J (2014) Transition management in five European cities – an evaluation. DRIFT, Erasmus University Rotterdam, Rotterdam

Roorda C, Hölscher K, Nevens F (2014a) The climate arena in Ghent. Working paper, DRIFT

Roorda C, Wittmayer J, Henneman J et al (2014b) Transition management in the urban context: guidance manual. DRIFT, Erasmus University Rotterdam

Rotmans J, Loorbach D (2011) Towards a better understanding of transitions and their governance: a systemic and reflexive approach. In: Grin J, Rotmans J, Schot J (eds) Transitions to sustainable development. New directions in the study of longer-term transformative change. Routledge, New York/London, pp 105–220

Swyngedouw E (2005) Governance innovation and the citizen: the Janus face of governance-beyond-the-state. Urban Stud 42(11):1991–2006. doi:10.1080/00420980500279869

Thomas KW, Velthouse BA (1990) Cognitive elements of empowerment: an 'interpretative' model of intrinsic task motivation. Acad Manag Rev 15(4):666–681

van den Bosch S (2010) Transition experiments: exploring societal changes towards sustainability. Dissertation, Erasmus University Rotterdam

Wittmayer J, Frantzeskaki N, Steenbergen F van et al (2011) Towards an urban change agents' analysis method: experiences from five European cities. Paper presented at Resilient Cities 2011, 2nd world congress on cities and adaptation to climate change, Bonn, 3–5 Jun 2011

Chapter 7
Case Study of Eco-town Project in Kitakyushu: Tension Among Incumbents and the Transition from Industrial City to Green City

Hideaki Shiroyama and Shinya Kajiki

Abstract In the decades following World War II, the city of Kitakyushu in southern Japan gained an international reputation for its steel production and industrial pollution. But in the early 1990s, the city government (hereafter, "City") began to change its industrial structure and pursue a course of sustainability through environmental conservation, the promotion of environmental business, and welfare for local citizens. The Kitakyushu Eco-town Project, an initiative aimed at promoting environmental business such as polyethylene terephthalate (PET) bottle recycling and home appliance recycling, was started in 1997 to effectively utilise reclaimed land in Hibikinada, located in the northwestern part of the city. The concept for utilising reclaimed land was discussed within the public sector (Committee for Hibikinada Basic Development Plan), and specific contents were conceived by the private sector (study group and advisory committee). Pressure from incumbent personnel, namely insiders with new ideas in established private and public institutions, was important in facilitating the transition. In a short period of time, the Eco-town has expanded domestically, both geographically and in terms of the scope of sectors represented, and overseas to China. The Kitakyushu Eco-town Project can be viewed as a catalyst of Kitakyushu's transition from an industrial city to a green city. The public–private collaborative networks and individual expertise developed through this project have had a longstanding influence on the design of subsequent projects such as Environmental Model City and Smart Community projects.

In this case study, the authors analyse, through document analysis and interview surveys, the transition process Kitakyushu followed from an industrial city with controlled pollution to a green city with a strong focus on the promotion of environmental industry and sustainability with multiple dimensions including environment, healthcare, and economic values.

H. Shiroyama (✉)
The University of Tokyo, Tokyo, Japan
e-mail: siroyama@j.u-tokyo.ac.jp

S. Kajiki
Graduate School of Public Policy, The University of Tokyo, Tokyo, Japan

D. Loorbach et al. (eds.), *Governance of Urban Sustainability Transitions*, Theory and Practice of Urban Sustainability Transitions,
DOI 10.1007/978-4-431-55426-4_7

Keywords Kitakyushu • Nippon steel • Eco-town Project • Ministry of Economy, Trade and Industry • Ministry of Environment • Environmental industry • Recycling • Incumbents

Main Actors and Organizations

City of Kitakyushu: A designated city with a population of 1 million people on Kyushu, Japan's southernmost main island.

Koichi Sueyoshi: City mayor from 1987 to 2007 (former bureaucrat of the Ministry of Construction).

Committee for Hibikinada Basic Development Plan (public sector): Set up by Mayor Sueyoshi in 1989 to plan utilization of Hibikinada reclaimed land in the City.

Nippon Steel (currently Nippon Steel & Sumitomo Metal Corporation): Steel manufacturer formed in 1970 through the merger of Fuji Iron & Steel and Yawata Iron & Steel. Based in Tokyo.

Nippon Steel Yawata (former Yawata Iron & Steel): Iron factory of Nippon steel located in the Yawata district of Kitakyushu.

Junichi Kawasaki: Executive in the general affairs department of Nippon Steel Yawata.

Yoichiro Kamei: President of Kyushu Kogyo Co., Ltd. (subsidiary of Mitsui & Co., Ltd.).

Study Group & Advisory Committee (private sector): Established by Nippon Steel Yawata and Mitsui & Co., Ltd. Consists of members of these companies and city officials for planning the utilisation of Hibikinada reclaimed land for new green business.

Eco-town Project: National project organised by the Ministry of Economy, Trade and Industry and the Ministry of Health and Welfare (later Ministry of the Environment) for local revitalisation through the promotion of green business.

Date	Event
October 1989	Committee for Hibikinada Basic Development Plan formed
November 1994	Study group formed by two company executives
March 1995	Nippon Steel Yawata and Mitsui & Co., Ltd. establish Advisory Committee (first phase) that includes city officials
June 1995	Law for Promotion of Sorted Collection and Recycling of Containers and Packaging was drafted by the Ministry of International Trade and Industry (MITI)

(continued)

Date	Event
February 1996	Nippon Steel Yawata and Mitsui & Co., Ltd. establish Advisory Committee (second phase) that includes high-ranking city officials
April 1997	Eco-town Project begun by the Ministry of International Trade and Industry and the Ministry of Health and Welfare
July 1997	Kitakyushu Eco-town Project (first stage) launched
August 1997	Kitakyushu Environmental Industry Promotion Council established
July 1998	PET bottle recycling business started in Kitakyushu
August 2002	Kitakyushu Eco-town Project (second stage) launched
July 2005	Kitakyushu Eco-complex Promotion Council formed
September 2007	Overseas expansion of Kitakyushu Eco-town Project begun

7.1 Background and Analytical Framework

7.1.1 History of Kitakyushu: Moving Beyond Pollution

Kitakyushu is located at the "entrance" to Kyushu in Japan and is a "designated city" (city with more than 500,000 people and commensurate administrative functions) with a population of 1 million people. The city's development began in the late nineteenth century. In 1901, the first modern steel plant in Japan, the government-operated Yahata steel works (currently Nippon Steel & Sumitomo Metal Corporation), began operations. This steel works, which became Japan's top steel supplier, was situated near the site of Japan's largest coalfield, and railroad lines and port facilities were also established. Taking advantage of these resources, the city thereafter developed essential industries such as chemical, metal, and ceramic industries, which had a significant role in Japanese modernization. The smoke from these factories' chimneys was a symbol of local prosperity.

However, air and water pollution associated with this industrial development soon became prevalent. By the late 1950s, environmental conditions in Kitakyushu were severely threatened. High concentrations of soot and dust caused NO_x, particulate matter, and SO_x levels to rise beyond World Health Organization (WHO) standards. Water pollution in Dokai Bay, at the heart of the industrial district, became so bad that it corroded screws on vessels and even killed *Escherichia coli* bacteria. The number of complaints about foul smells rose. Kitakyushu moved to the forefront of the unsustainability problem arising from Japan's modernization.

To reduce this industrial pollution, women in the city organised. The association delivered petitions and strongly demanded improvements to the companies based on the results of their own investigation. They also produced a movie titled "Aozora ga Hoshii" ("We Want Blue Skies"), which revealed the acuteness of the problem for the rest of the country.

Against this backdrop, the City created a new department in charge of pollution control in 1963 and raised it to the status of Pollution Control Bureau in 1971. In the

1960s and 1970s, it commissioned experts in various fields to study the structure, operations, and maintenance of smoke-emitting plant facilities and precipitators and discussed various feasible countermeasures. From 1969 to 1972, the City signed "pollution control agreements," informal voluntary agreements outlining specific pollution control measures, with each factory. In addition, companies reduced emissions by retrofitting their facilities with newer and higher-efficiency equipment.

As a result of having taken these measures through civic pressure and the City's own administrative efforts, companies made significant progress in developing new technologies and improving the production process. Many pollution issues were settled by the late 1970s.

According to Kishimoto (2011), these antipollution measures have been dubbed the "Kitakyushu method," of which there are three characteristics:

1. Close partnership among citizens, local government, and companies
2. Positive initiative by local government
3. Sustained private-sector technology development and deployment

Once pollution issues were resolved, the search for international collaboration in environmental and engineering fields began. In 1980, the Kitakyushu International Training Association (KITA[1]) was established through partnership between the City, Fukuoka Prefecture, Junior Chamber International Kitakyushu,[2] the Kitakyushu Chamber of Commerce, and other organizations. In the same year, KITA started a Japan International Cooperation Agency (JICA) training course to contribute to international cooperation. In addition, the City began in the 1980s to work on environmental international collaboration with cities in China and other parts of Asia. For example, the City gave a lecture on pollution management in Dalian, China, in 1981 and began providing assistance on a plan to develop Dalian as a model environmental city using Official Development Assistance (ODA) resources in 1993; a joint investigation by the City and JICA suggested general environmental measures to incorporate into the city's plans.

The oil crisis in the late 1970s, appreciation of the yen, and structural economic change in the 1980s had a large influence on Kitakyushu's economy. Impacts on the iron and steel industry were particularly great. Mayor Sueyoshi, who was elected in February 1987, developed a "Kitakyushu Renaissance Plan" based on his pledge to revitalise the regional economy. The basic concept of this plan was "to become an international and technological city with waterfronts, a green environment, and human connection," which included, as a component, the "promotion of environmental industries" through the utilization of the reclaimed land in Hibikinada, located in the northwestern part of the city. He worked on various projects during

[1] KITA was renamed Kitakyushu International Techno-cooperative Association (KITA) in 1992.

[2] Junior Chamber International (JCI) is a nonprofit organization of young active citizens who are engaged in and committed to creating impact in their communities.

his 20 years (until 2007) in office, and "Kitakyushu Eco-town Project" is a representative example.

Movement toward creating a more sustainable society began in the latter half of the 2000s throughout Japan. Kitakyushu was selected as a "Eco-model city" by the Cabinet Office in July 2008, as a "Next-Generation Energy Infrastructure and Social System Demonstration Area" by the Ministry of Economy, Trade and Industry (METI) in April 2010, and as a "future city" by the Cabinet Office in December 2011, based on its long experience in environmental management.

In addition, the City has built cooperative frameworks with other cities in Asia, such as the Asian Environmental Cooperation City network and an Environment Committee in the Organization for the Promotion of East Asia Economic Development. Companies and academia are also working on numerous projects with the City at the Kitakyushu Asian Center for Low Carbon Society, established in 2010.

In June 2011, the Organization for Economic Co-operation and Development (OECD) selected Kitakyushu as the first "green growth demonstration city" in Asia. The OECD's working paper "Green Growth in Kitakyushu, Japan" observes that civic initiative and dialogue between various stakeholders propelled the switch from a "grey" city to a "green growth" city. This report also states that these experiences have been utilised in various subsequent projects (OECD 2013). The Kitakyushu Eco-town Project is a representative example.

7.1.2 Using Transition Management as an Analytical Framework

In this chapter, the authors analyse the process of the Kitakyushu Eco-town Project, focusing on the PET bottle recycling business that had a key role in this project, as a catalyst of transition from an industrial to a green city.

The Kitakyushu Eco-town Project has various objectives that differ from the antipollution measures of the 1960s and 1970s, including the utilization of reclaimed land and the creation of new recycling businesses. Moreover, both the public–private collaborative networks and individual expertise developed through this project have had a longstanding influence on the design of subsequent projects. For example, the Conference for Kitakyushu Smart Community Project (Next-Generation Energy Infrastructure and Social System Demonstration Area adopted by METI in 2010), which aims to develop and internationally expand the Japanese version of a smart grid, also consists of public- and private-sector stakeholders, such as the City, Nippon Steel, Mitsui & Co., Ltd., IBM Japan, Toyota Motor Corporation, and Kyushu Electric Power Company. In addition, key figures, such as Taihei Shibata, director of the City's Smart Community Section, were trained in management of the Eco-town Project starting in 1997.

To analyse the long-term development of the Kitakyushu Eco-town Project, the authors use a "transition management" approach as a framework for analysing a historical transition. Transition management is a governance approach to sustainability transition. It is based on an understanding of transition as processes of fundamental long-term multilevel and multiphase changes in complex, adaptive systems. In our understanding, the Kitakyushu Eco-town Project aimed at fostering a transition to a more sustainable city through revitalization and the promotion of green businesses. As such it makes sense to analyse it using a transition management perspective. The framework of transition management can be operationalised into different governance instruments such as transition arena and transition experiments. It can also be operationalised into more detailed process methodologies guiding actors in implementing transition governance activities for orienting, agenda setting, activating, and reflecting (Wittmayer and Loorbach 2016, Chap. 2, this volume). In this chapter, we use a transition management lens for analysing governance dynamics in the urban context of Kitakyushu.

This chapter focuses on the initial phases of transition of Kitakyushu from industrial to green city. In other words, this chapter focuses on the roles of the "transition team" in addition to the "transition arena." The transition team is a core team that adapts and drives the transition management process (Roorda et al. 2014, p. 47); a transition arena, however, is a setting that provides an informal but well-structured space for a group of change agents to critically reflect on the status quo and develop alternative ideas, practises, and relations (Roorda et al. 2014, p. 46) (Fig. 7.1).

By using these frameworks, the authors focus on the following five activities.

1. *Orienting*: The process initiated by transition team and transition arena (formal process in the public sector and informal process initiated by private company executives)
2. *Agenda Setting*: Developing an agenda (alignment with national legislation and negotiation with corporate leaders in headquarters)
3. *Activating 1*: Experiments in this project (PET bottle recycling, first and second stages of the Kitakyushu Eco-town Project)
4. *Activating 2*: Expansion at home and overseas (Japan, overall city area; overseas, Qingdao, Tianjin, Dalian)
5. *Reflecting*: Learning lessons from this project

7.2 Exploring Ideas on Utilization of Hibikinada

7.2.1 Formal Process in the Public Sector

In the 1980s, a very large swath of land (2000 ha) was developed in the Hibikinada district located in the northwestern part of the Kitakyushu because the City and companies had reclaimed the foreshore with sand and slag. However, the original plans for industrial development were thwarted by structural changes in the heavy manufacturing industry, and the City was forced to rethink its plans. In October

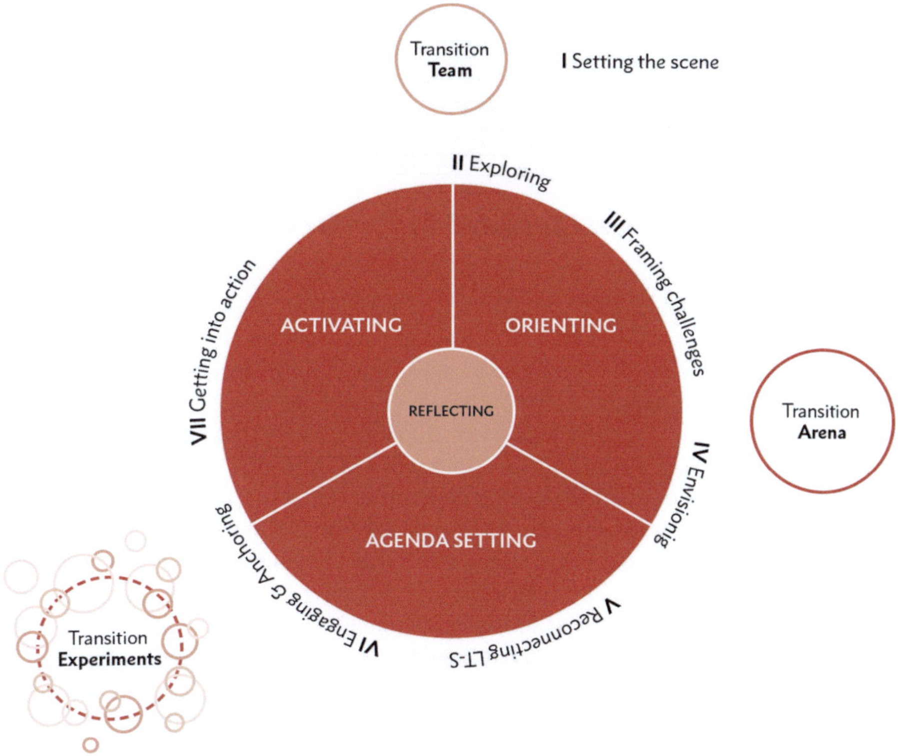

Fig. 7.1 The transition management process structure (From Roorda et al. 2014, p. 14)

1989, the City set up a "Committee for Hibikinada Basic Development Plan" (chairman: Prof. Toshifumi Yata of Kyushu University), which included representatives from Nippon Steel, Mitsubishi Chemical, Mitsui & Co., Ltd., and Hibikinada Kaihatsu Co., Ltd., to consider the possible uses of the Hibikinada district. In March 1992, the committee agreed on the "Hibikinada Basic Development Policy" for fostering the waste management and recycling industries ("venous industry") utilizing technologies and human resources based on Kitakyushu's long experience in pollution management. In line with this Basic Policy, the City began holding study sessions with relevant bureaus. Finally, in March 1996, the Committee formulated the Hibikinada Basic Development Plan, which consisted of the following four items.

1. Attract and support environmental industries
2. Establish a R&D base for environment-related fields
3. Improve infrastructure for development and supply of energy
4. Attract and support industries relating to environmental devices and engineering

Preceding this policy was organisational change in the municipality. The Pollution Control Bureau merged with the Environment Operation Bureau in charge of waste management, creating in 1990 a new Environment Bureau in charge of environmental policy, waste management policy, and industrial policy (City of Kitakyushu 1998, p. 105). This bureau was placed in charge of the Hibikinada district development and the Eco-town Project.

7.2.2 *Informal Process Initiated by Private Company Executives*

Around the same time, amidst a downturn in iron manufacturing, Nippon Steel (currently Nippon Steel & Sumitomo Metal Corporation), which owned 300 ha of unused land in Hibikinada district, started discussions on the creation of new business utilizing this land. In June 1994, Junichi Kawasaki, who belonged to the iron manufacture section of Nippon Steel Yawata, moved to the general affairs department and started working on the creation of new business. Although IT and the environment were key business themes at the time, Kawasaki chose the latter and set up a study group in November 1994 with Yoichiro Kamei, who was the president of Kyushu Industry (subsidiary of Mitsui & Co., Ltd.) and knew the recycling business well. Kawasaki and Kamei knew each other well because their children were going to the same school. In the study group, it became clear that it would be very difficult to create a waste recycling business, but they continued to search for the possibility of some form of waste recycling business because the City had been considering promoting the environmental industry in the Committee for Hibikinada Basic Development Plan.[3] These two were incumbent members of companies but were trying to introduce new ideas.

Kawasaki and Kamei expected it would be important to understand the City's policy on waste management administration, and so in March 1995 they set up an advisory committee (first phase) composed of members of the private sector (Nippon Steel Yawata, Nippon Steel Engineering, Mitsui & Co., Ltd., Hibikinada Kaihatsu) and city officials from various bureaus and departments. Similar public–private cooperative networks were also constructed in future projects, such as the Smart Community Project (Next-Generation Energy Infrastructure and Social System Demonstration Area).

Because most members of the advisory committee from the City (planning bureau, environmental bureau, economic affairs bureau, port and harbour authority, and city planning departments) were in charge of the secretariat of the City's Committee for Hibikinada Basic Development Plan, it was easy to keep track of their deliberations. Based on the extensive experience of its members, the advisory committee concluded that a new environmental industry, dissimilar from the conventional waste disposal treatment business, would need to be created.

[3] Interview with Junichi Kawasaki, former executive of the general affairs department of Nippon Steel Yawata, on June 26th, 2013 in Kitakyushu city.

In the first phase of the advisory committee, participating city officials were those at the rank of subsection chief. However, because political support from the City was critical to the success of the recycling business, Kawasaki and Kamei asked for the attendance of officials at the level of section manager or director in the second phase of the advisory committee from February 1996. The advisory committee developed a Comprehensive Environmental Industrial Complex concept that would later become the starting point for the Eco-town Project.

In a municipal council meeting in February of the same year, the question was posed to Mayor Sueyoshi, "Do you intend to make the city a dump site?" His answer: "These are resources for new industry, not garbage."[4] From this anecdote, it is clear that the creation of an environmental industry had become an important City policy.

7.2.3 Alignment with National Legislation and Negotiation with Corporate Leaders in Headquarters

After these advisory committee meetings, Kawasaki and Kamei started examining the idea of a PET bottle recycling business based on the Law for Promotion of Sorted Collection and Recycling of Containers and Packaging (Containers and Packaging Recycling Law). This legislation was drafted by the Ministry of International Trade and Industry (MITI) in June 1995, coincidentally around the same time when the idea for PET bottle recycling was put on the agenda in Kitakyushu. The law became the decisive factor in choosing the type of business to create, because the law clarified the roles of consumers (classification), municipalities (separated collection, storage), and the company (recycling).[5]

However, negotiating with Nippon Steel leaders in headquarters (located in Tokyo), who had the authority to decide, was not easy. Executives at the head office were concerned about the difficulty of bearing factory construction costs of 1.6 billion yen. In addition, they pointed out the contradiction of working in a competing field such as the PET bottle recycling business, because the company had a division that produced and recycled steel cans. After several stages of negotiation between the vice-president of the head office and the Nippon Steel Yawata iron manufacture director, the problem was finally settled as a "land sale item," taking into account also the history of the company's collaborative relationship with the City.

How to frame the issue was an important part of obtaining agreement from the corporate leaders at headquarters. The company was restructuring at the time, and it became a problem to let Nippon Steel Yawata that had made an effort for reduction

[4] Interview with Satoshi Nakazono, former manager of environmental conservation at the City of Kitakyushu, on June 26th, 2013 in Kitakyushu city.

[5] Interview with Kawasaki on June 26th, 2013 in Kitakyushu city.

in cost bear a large amount of property tax imposed on the land in the Hibikinada district, and the Nippon Steel head office made a plan to bear it in a lump in substitution. One idea that emerged to persuade leaders at headquarters was to sell the land to launch the PET bottle recycling business. Under the devised scheme, the PET bottle recycling company would purchase the land from Nippon Steel headquarters for 300 million yen, thereby decreasing its financial burden. This step would also transfer responsibility for the result of the business decision away from the company as a whole to Nippon Steel Yawata. To this, headquarters could agree. Thus, the proprietary rights of the unused land shifted from headquarters to Nippon Steel Yawata.

Through these processes, West Japan PET Bottle Recycling Co., Ltd. (NPR) was established in April 1997 based on investments from five private enterprises (Nippon Steel, Mitsui & Co., Ltd., Nittetsu Transportation, Nippon Express, and Sankyu) and the City. The new venture became the first in the Eco-town Project designated by central government (MITI and the Ministry of Health and Welfare) in July, as is explained later.

7.3 Eco-town Project: Interactions with National Policy Developments

7.3.1 National Policy Developments

Because Kitakyushu City actively shared personnel with the central government, it was easy to exchange information and opinions with the central government ministries. When Kazuo Matsunaga, the section manager of environmental policy at MITI, visited the City in May 1996 looking for new policy ideas, the City suggested that establishing new business would require differentiating the recycling business from conventional garbage collecting.[6] The idea of completely eliminating waste and emissions by utilizing them as resources was attracting attention at the time, and the Ministry and City initially made a plan for a business creation project that would have the central government providing 5 % and the city providing 2.5 % financial support.

However, negotiations between MITI and the Ministry of Finance resulted in the creation of the Eco-town Project, a national program aimed at revitalising local economies through the promotion of green business (environmental industry) and the creation of an environmentally harmonious system by industry, the public sector, and consumers. Under this new framework, up to 50 % of the project cost would be borne by the central government, provided that MITI and MHW (subsequently the Ministry of the Environment) approved the plan that the local government devised. It was a very attractive subsidy scheme.

[6] Interview with Satoshi Nakazono on June 26th, 2013 in Kitakyushu city.

7.3.2 Strategy of Kitakyushu City Government and Training of Officials as Transition Managers

Hearing of this new opportunity, the City prepared a plan for the Kitakyushu Eco-town Project based on various existing plans, and West Japan PET Bottle Recycling Co., Ltd. (NPR) was approved as the first plant in the new "Eco-town" subsidy scheme. Two types of subsidies were provided in this scheme to stimulate environmental industries: a "soft" subsidy for project preparation costs and a "hard" subsidy for infrastructure improvement costs. Because Kitakyushu City was the only municipality to prepare a concrete plan in the first year, the entire hard subsidy was allotted to construction of the PET bottle recycling plant. The budget was significantly increased in the second year; numerous recycling plants for office equipment, automobiles, home appliances, fluorescent lights, and the like were established in the Hibikinada district under this scheme.

This strategy, that is, planning an Eco-town Project before the central government and becoming the first local government to receive assistance for a project under the new subsidy scheme, was made possible by the wealth of experience of Mayor Sueyoshi, who worked in the Ministry of Construction for about 30 years. The mayor knew well the budget decision-making process at the ministries and central government offices and wrote the reasons why the City was able to apply for Eco-town Project funding as follows (Abe and Inaba 2013, pp. 240–241).

> Each ministry and government office begins preparing policy and budget of the following fiscal year around June when ordinary session of the Diet and yearly personnel transfer have finished. In fact, this is the espionage period for local governments. For example, policies that were not adopted this year may be adopted the next fiscal year. Collecting such information and making business plans for next fiscal year is important. The existence of local governments that can utilize the remaining budget is also fortunate for the central government. After the central government distributes its budgets for the year, because of the circumstances in each area, some projects may not progress well and it might be necessary to distribute the remaining resources to other areas. Kitakyushu Eco-town Project also utilized this additional distribution. I always prepared around 50 plans, everything from public works to welfare, to education projects. When it looked like additional budget might be distributed to a project such as the Eco-town Project, I appealed to each ministry and government office.

Multiple ministries and government offices, including MITI, the Ministry of Construction, and the Ministry of Education, had jurisdiction over the recycling business. So Mayor Sueyoshi devised a strategy to draw subsidies from the various ministries and government offices in every project. For this he put together a team of subsection chief-level officials from various fields (civil engineering, chemical engineering, mechanical engineering, biotechnology, agriculture, law, and business administration) to negotiate with the ministries (Abe and Inaba 2013, pp. 240–241). Now, most of these officials are at an executive level and lead various projects such as the "Smart Community" project (Next-Generation Energy Infrastructure and Social System Demonstration Area Project) and serve as mediators between local government and central government.

Furthermore, the City adopted a "catered lecture" program in 1989, requiring officials in executive-level positions to explain city plans or projects beyond their jurisdiction to citizen groups composed of more than 20 people. This program was an opportunity to hear public opinion and also a venue where such officials could be tested on their ability to respond to input. It was also an opportunity for officials to develop a capacity to communicate with various stakeholders. In other words, this program was a mechanism for training future transition managers. Such a "catered lecture" was held for the PET bottle recycling business. Residents expressed anxiety over how garbage would be collected, the potential rise in illegal dumping, and the possibility of Hibikinada simply becoming a dump site.[7] To obtain residents' consent, City officials, with support from Nippon Steel Yawata, had to explain that "the Environmental Bureau does not take part in environmental pollution" and "a certain local trusted company will take responsibility."

In addition, an employee of NPR and officials of the Environmental Bureau called on other local governments in Kyushu for help in collecting PET bottles, because 8000 t of bottles per year was calculated as necessary for the business to make a profit.

7.4 Developments of Kitakyushu Eco-town Project

7.4.1 First Stage (1997–2002): Development of Recycling Business

The Kitakyushu Eco-town Project began in July 1997. The newly formed Kitakyushu Environmental Industry Promotion Council, composed of Nippon Steel Yawata, Mitsui & Co., Ltd., Toshiba, Hitachi, academia, the chief of the Kyushu Economic Industrial Bureau, and the deputy-governor of Fukuoka Prefecture, formulated three action plans, one each for education and basic research, technology and demonstration studies, and industrialisation, as a three-pronged strategy to environmental industrial development.

Research institutions such as Kitakyushu Municipal University, the Kyushu Institute of Technology, Fukuoka University, and Waseda University gathered at the Kitakyushu Science and Research Park that was established in April 2001 to participate in the education and basic research area. In the technology and demonstration studies area, various research projects on technologies relating to the environment and recycling were undertaken under a framework of collaboration between companies, universities, and governments.

In the industrialisation area, companies began commercializing recycling operations, the starting point of the envisioned Comprehensive Environmental Industrial Complex. This was followed by a flurry of activity from various businesses

[7] Interview with Satoshi Nakazono on June 26th, 2013 in Kitakyushu city.

aimed at contributing to the creation of a zero-waste industry hub. The factory at NPR, the company that played a key role in launching the Kitakyushu Eco-town Project, started operations in July 1998, producing egg cartons and seats from resin recycled from the chemical fibre of PET bottles.

From there, six recycling plants for such things as office automation equipment, automobiles, home appliances, fluorescent lamps, and medical equipment were built in the area. The Home Appliance Recycling Law, which was enacted in June 1998, catalyzed the creation of Nishinihon Kaden Recycle Corporation (NKRC) in the same year by financing from manufacturers.

The recycling of automobile waste, meanwhile, was made possible by methods developed by Yoshikawa Industry Co., Ltd. (Partner Company of Nippon Steel) and West Japan Auto Recycling Co., Ltd. Financed by Nippon Steel, Mitsui & Co., Ltd., a new company started operation in February 2000. Furthermore, to support small and medium-sized enterprises and start-up companies in the recycling business, the City established a "frontier zone" and "automobile recycling zone" where it lent land with fixed-term lease rights at Hibiki Recycling Park adjacent to the Comprehensive Environmental Industrial Complex.

In this way, recycling business in the Kitakyushu Eco-town Project started with PET bottles and expanded to other fields in what became the first stage (1997–2002).

7.4.2 *Second Stage (2002–2010): Establishing Mutual Cooperation with Broader Recycling-Related Companies in "Eco-town"*

Most businesses in the first stage of the Kitakyushu Eco-town Project Plan were realised ahead of schedule. By the time the second stage began in August 2002, Japan's Basic Law for Promoting the Creation of a Recycling-Oriented Society had already come into effect (January 2001) and greatly affected social conditions.

The Kitakyushu Eco-town Project Plan called for the business area to be expanded to the eastern whole of Hibikinada district in the second stage. As Fig. 7.2 shows, the number of businesses in Kitakyushu Eco-town has increased steadily since the second stage began. Today, Kitakyushu Eco-town is composed of three areas, that is, the Kitakyushu Science and Research Park, the Technology and Demonstration Studies Area, and the Comprehensive Environmental Industrial Complex (Hibiki Recycling Park). In the Demonstration Studies Area, a waste research facility was constructed in February 2003, and studies on topics such as rapid stabilization and management of the landfill sites and contaminated soil treatment technology have been conducted. Furthermore, a subsidy scheme for development of advanced environmental technologies was established in April of the same year to support environmental fields that were mature and commercially feasible. Seven companies that were established to solve problems such as noise

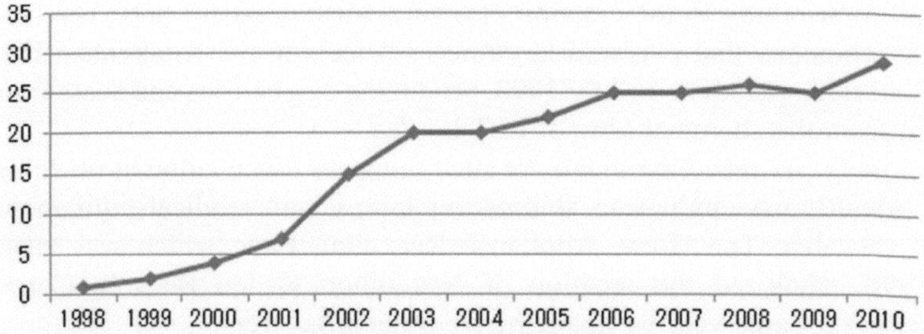

Fig. 7.2 Change in the number of businesses in Kitakyushu eco-town (From Ministry of the Environment 2012)

and traffic obstruction/congestion in the city area moved to Hibiki Recycling Park, thereby forming a large automobile recycling base.

Companies that have moved to Eco-town have made an effort to collaborate with other companies by using their by-products. Nonetheless, the use of intermediate waste treatment facilities or landfill sites outside the area is still necessary, because recycling all by-products in Eco-town is difficult for economic or technical reasons. To address this problem, a multifunctional central facility for incinerating by-products and recycling residual heat was built in 2005.

7.4.3 Expansion at Home and Overseas

Allowing companies in the project to select the most appropriate places in the city to operate their unique businesses necessitated the expansion of the project's area. So, the City applied to the central government to expand the project area to the entire city, which was approved in October 2004.

In July 2005, the Kitakyushu Eco-Complex Promotion Council (composed of 17 private enterprises and four universities, the chief of the Kyushu Economic Industrial Bureau, the deputy-governor of Fukuoka Prefecture, and the City) was also formed starting the move to optimise resources and energy use between the industrial and residential sectors on a regional level.

In addition to expanding the project area to the entire city, the City also started examining the possibility of developing business overseas. The main concept of the Kitakyushu Renaissance Plan developed in 1987 was to "become an international and technological city with waterfronts, a green environment, and human connection." From the latter half of the 1990s, Mayor Sueyoshi had a plan to contribute to the solution of global environmental problems through international cooperation, and expected that such cooperation would promote economic relations between cities, including the economic growth of Kitakyushu over the medium- to long term (Hashiyama and Arata 2010, p. 150).

Table 7.1 Eco-towns in Japan (1997–2005)

Fiscal year	Area
1997	Iida City, Kawasaki City, Kitakyushu City, Gifu Prefecture
1998	Ohmuta City, Sapporo City, Chiba City/Chiba Prefecture
1999	Akita Prefecture, Uguisuzawa Town
2000	Hokkaido, Hiroshima Prefecture, Kochi City, Minamata City
2001	Yamaguchi Prefecture, Naoshima Town
2002	Toyama City, Aomori Prefecture
2003	Hyogo Prefecture, Tokyo, Okayama Prefecture
2004	Kamaishi Town, Aichi Prefecture, Suzuka Prefecture
2005	Osaka Prefecture, Yokkaichi City, Ehime Prefecture

Source: Compiled from the published data of the Ministry of the Environment

Following this idea, the City set a goal in the second stage of the Kitakyushu Eco-town Project Plan to promote the 3Rs and establish an environmental industry base city in Asia beyond Japan through systematic accumulation of various industries related to recycling resources and technologies and further expansion of previous actions. Although the Japanese government (METI and Ministry of the Environment) acknowledged the Eco-town Project as a means of local revitalization and approved 26 areas across Japan (Table 7.1), the City went further and attempted to spur future economic growth through international cooperation.

It becomes clear at this point that the goal of the Kitakyushu Eco-town Project grew from the effective utilization of Hibikinada district to include international cooperation in the second stage.

The City targeted places in China through sister-city agreements and the Organization for the Promotion of East Asia Economic Development. In 2007, under a bilateral agreement between Japan and China, three Chinese cities (Qingdao, Tianjin, and Dalian) were selected as the location of Japan–China cooperation projects, in which the City of Kitakyushu has been involved. For example, the following five specific items were adopted in Qingdao.

1. Cooperation on a "recycled resources industry development plan" that Qingdao City devises
2. Examination of the possibility of introducing technologies and facilities for collecting and processing home appliance waste
3. Consideration of the potential for cooperation with existing or planned recycling-related industries in Qingdao
4. Visiting Japan to train city officials and relevant parties from companies in Qingdao
5. Advertising this cooperation project in both Japan and China

The overseas expansion of the "eco-town" concept and the Kitakyushu City business model have included not only technology transfer but also legislation and operational management of waste collection systems. The City aims to integrate these areas, developing everything from start to finish, including the construction of institutional systems and the creation of recycling markets.

In addition, improvements in waste collection methods are indispensable to improving the operation of recycling machinery as well as to introducing technology that allows Eco-town businesses to secure a profit.[8] Training of operating staff is therefore necessary. The City has also been matching Kitakyushu companies with companies overseas to support their foreign operations.

In this way, Kitakyushu City has expanded the geographic scope of its Eco-town to the entire city, and even transferred technologies and operational systems to China through international cooperation networks.

7.5 Concluding Remarks

7.5.1 Kitakyushu Case Analysed from a "Transition Management" Framework

As described here, the Kitakyushu Eco-town Project started as an exploration into the effective utilisation of unused land in the Hibikinada district, with involvement from a variety of interested parties including both the public sector and private sector. The study group and advisory committees set up by Nippon Steel Yawata and Mitsui & Co., Ltd. examined more concrete actions, and Kitakyushu City's Committee for Hibikinada Basic Development Plan discussed the project on a general conceptual level.

The PET bottle recycling business that key private-sector figures proposed was later adopted as an Eco-town Project by the central government. From there, the geographic and industrial scope of recycling in the Eco-town expanded, facilitated by the Home Appliance Recycling Law and, later, the Basic Law for Promoting the Creation of a Recycling-Oriented Society. Because Kitakyushu Eco-town was a project originally aiming at effective utilisation of Hibikinada, expansion was not planned in the first stage (1997–2002). However, examination of the possibility of expanding the project to the entire city and to areas beyond, such as China, started in the second stage (2002–present). Through these "transition experiments," the City achieved transition from an industrial city with controlled pollution to a green city focusing on the promotion of integrated environmental industries.

Figure 7.3 shows the process of the Kitakyushu Eco-town Project's development, and Table 7.2 identifies the major instruments used in this process, such as transition team, transition arenas, and transition experiments based on Roorda et al. (2014, pp. 14–15). In the case of the Kitakyushu Eco-town Project, the transition team was composed of key members of the study group launched by

[8] Interview with Kengo Ishida, executive director of the Environment Bureau of the City of Kitakyushu, on June 26th, 2013 in Kitakyushu city.

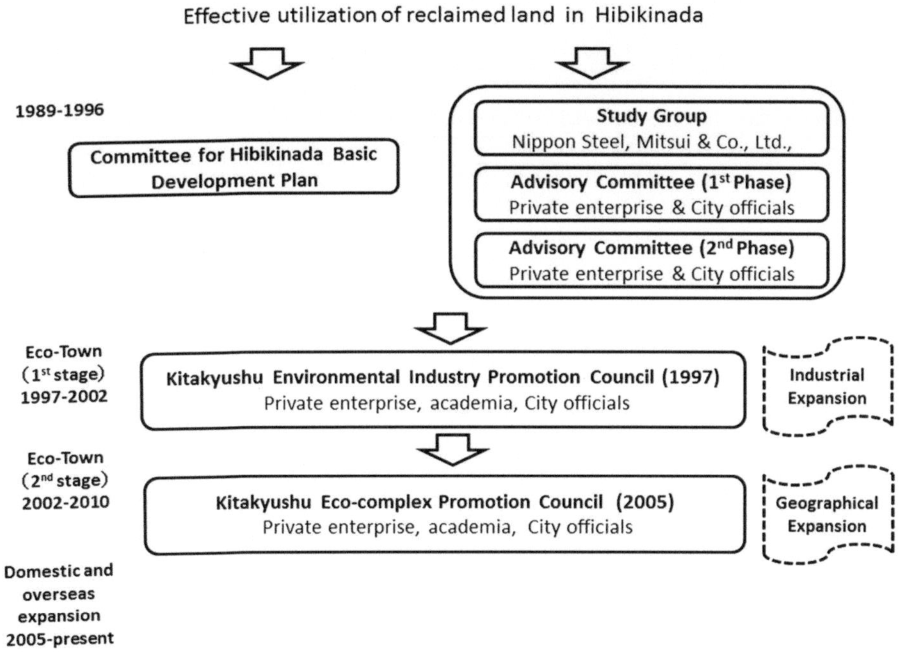

Fig. 7.3 Process of development of the Kitakyushu Eco-town Project

Nippon Steel Yawata and Mitsui & Co., Ltd. at a non-governmental level. The transition arenas consisted of key members of the Committee for Hibikinada Basic Development Plan set up by the City, and members of the advisory committees formed by Nippon Steel Yawata and Mitsui & Co. at a non-governmental level. Key figures, especially in the City government, overlapped with participants in those committees. Their members began to develop the idea of an "eco-town" beginning with the discussion of specific actions related to the effective utilisation of Hibikinada district.

Concrete project content was originally developed by actors in the private sector. Those ideas became the starting point for the development of Kitakyushu Eco-town. Members of the transition teams and transition arenas also performed the profitability assessment that led to creation of the PET bottle recycling business that became the first "transition experiment." City officials, who were participants in the "transition arena," had important roles in building the system to support this business through interactions with the central government. Relationships between the two advisory committees and the City government were mutually complementing. Participants of both committees from the City were executive-level officials who planned and led various subsequent projects in which a variety of interested parties now participate. These committees also provided an opportunity for officials to train and develop human networks.

The Kitakyushu Environmental Industry Promotion council established in 1997 and the Kitakyushu Eco-Complex Promotion Council established in 2005 were

Table 7.2 Major instruments of Kitakyushu Eco-town Project

Major instruments	Organizations and actions in Kitakyushu
Transition teams	Study group (Nippon Steel Yawata & Mitsui & Co., Ltd.)
Transition arenas	Advisory committee (first phase) Advisory committee (second phase) Committee for Hibikinada Basic Development Plan
Transition experiments	Transition experiments (first stage) PET bottle recycling business Transition experiments (second stage) Industrial expansion: expansion of scope of recycling industries (automobile recycling, home appliance recycling, etc.) Geographic expansion: all of Kitakyushu, and China (Qingdao, Tianjin, Dalian)

used at the implementation stage in which a broader variety of interested parties were involved and discussed concrete plans for "transition experiments." These councils both worked to expand their treatment fields and area. Expansion of the Kitakyushu Eco-town Project has followed two types: industrial expansion (from PET bottle recycling to a broader recycling industrial base) and geographic expansion (from Hibikinada district to all of Kiakyushu, and to China). In the Kitakyushu Eco-town Project, home appliance and automobile recycling businesses relocated to Hibikinada following the PET bottle recycling business, turning the district into a recycling industrial base. Kitakyushu Eco-town eventually expanded its reach to the entire city. The City then attempted overseas expansion through established sister-city networks and by spearheading Japan–China cooperation projects with three Chinese cities. International cooperation requires not only technology transfer but also the construction of institutional systems, the creation of recycling markets, and personnel training. Therefore, it is not easy to develop another "Kitakyushu Eco-town" simply through export.

7.5.2 Four Lessons from the Experience of Kitakyushu

There are four lessons we can learn from the experience of Kitakyushu's transition from an industrial city to a green city.

First, members of the private sector had crucial roles as key figures in the transition teams and transition arenas. Members of both the private and public sectors were incumbents, but they tried to introduce new ideas to change the existing regime. It can be said that pressure from these incumbents was the key driver of transition. The Committee for Hibikinada Basic Development Plan was already set up in 1989, but concrete ideas for a recycling business emerged only after these ideas were generated in discussions in the study group and advisory committees between 1994 and 1996, mainly by two major corporate figures, then brought to the formal arena of the Committee for Hibikinada Basic Development Plan. Informal discussion at the study group and advisory committees had an advantage in envisioning and generating concrete ideas. In addition, the close partnership between local government and companies formed during the antipollution era has been handed down as an important asset to the present. From this it becomes clear that selecting suitable talent is crucially important for forming a workable "transition team" and "transition arena." Furthermore, the special characteristics of the city must be given attention.

Second, that the private sector had an important role does not necessarily mean that public officials did not. City officials were important in bridging the promotion of environmental industries under Kyoto Protocol mechanisms at the national level and the generation of concrete ideas for recycling industry at the local niche level. These cases illustrates the potential of government actors as transition managers if they have sufficient connection with outside actors with new ideas. The introduction of the idea to set up a recycling plant when a senior official in charge of environmental policy at MITI visited the city was a turning point in the initial process of transition.

Third, one key ingredient in the role of public officials to bridge various levels and sectors of green city promotion was the training of officials as transition managers by the City through on-the-job experience. The City facilitated this training through "catered lectures," the sharing of personnel with central government ministries, and interaction with private-sector initiatives. Since 1989, executive-level officials have been required to give "catered lectures" in which they explain a City plan or project beyond their jurisdiction to an audience of ordinary citizens. This practise helps to improve officials' ability to think larger and to respond to various requests from citizens. In addition, the City actively shares personnel with central government ministries with aim of exchanging information, boosting personnel's planning skills, and building human networks. For example, Toshikazu Matsuoka, the current director of the Environmental Bureau who directs many projects, was transferred to the Ministry of the Environment from July 1989 to May 1992. Furthermore, as the advisory committees in this case show, public officials have been actively participating in informal discussions with the private sector. To achieve sustained "transition" success, the training of officials and personnel exchange in various fields will continue to be a crucial element.

Fourth, the Kitakyushu Eco-town Project was successful in expanding the scope of the local recycling industry from PET bottles to home appliances, to automobiles, and to various other sectors. The city has been successful in attracting other types of green business as well. The City has also formed networks with other cities involved in Eco-town projects and expanded the model overseas, such as to China. This transition process was not necessarily based on a grand vision at the initial stage: it was more of an incremental learning process that synchronised with transitions at the national government level, officials of which sometimes had an interest in using municipalities as a test bed for national transition.

References

Abe T, Inaba M (2013) Manuscript of Kouichi Sueyoshi. Nishinihon Shinbunsha, Fukuoka (in Japanese)

City of Kitakyushu (1998) History of anti-pollution measures in the city of Kitakyushu. City of Kitakyushu, Kitakyushu (in Japanese)

Hashiyama Y, Arata H (2010) Revitalizing Kitakyushu city as mayor. PHP Publisher, Kyoto (in Japanese)

Kishimoto C (2011) Creation of a sustainable local society in Kitakyushu after World War II. Perspect East Asia 22(1):23–36 (in Japanese)

Ministry of the Environment (2012) Working paper on effective collaboration policies between arterial and venous industries contributing to highly efficient utilization of existing venous industry hubs. Ministry of the Environment, Tokyo (in Japanese)

OECD (2013) Green growth in Kitakyushu, Japan. OECD, Paris

Roorda C, Wittmayer J, Henneman P, van Steenbergen F, Frantzeskaki N, Loorbach D (2014) Transition management in the urban context: guidance manual. DRIFT, Erasmus University Rotterdam, Rotterdam

Wittmayer JM, Loorbach D (2016) Governing transitions in cities: fostering alternative ideas, practices and actors, Chapter 2. In: Loorbach D, Wittmayer J, Shiroyama H, Fujino J, Mizuguchi S (eds) Governance of urban sustainability transitions, pp 17–36. Springer, Tokyo (this volume)

Chapter 8
Transition Management in Montreuil: Towards Perspectives of Hybridisation Between 'Top-Down' and 'Bottom-Up' Transitions

Adrien Krauz

Abstract The French government addressed the question of sustainability transitions by organizing a national debate on energy transition at the end of 2012, driven by the Ministry of Ecology. The transition motif has also been taken up by local and grassroots actors and initiatives trying to converge in a cohesive society project. These two strands of transitions, one institutional, directed in a 'top-down' manner, and another, a grassroots or 'bottom-up' sort, question the potential for a dialogue to emerge between these two dynamics and their respective actors.

This chapter draws on the case of Montreuil, the fourth most populous suburb of Paris, involved in a transition management process from 2011 to 2014 as part of the elaboration of its Local Climate Plan. Based on a 2-years' participant observation within the project team, the aim is to describe the adaptation of transition management to the French political context and to explore the possibility of 'hybridisations' between 'regime' and 'niche' actors—between their roles, relations, and cultures.

Keywords Transition • Sustainability • Transition management • Hybridisation • Montreuil • Reflexive governance • Civil servants • Empowerment

8.1 Introduction

8.1.1 'Bottom-Up' or 'Top-Down' Transitions?

The term 'transition' (under the labels 'post-carbon transition,' 'energy transition,' 'ecological transition') is emerging in France in reference to a state-driven change

---

A. Krauz (✉)
Paris West University, Mosaïques-Lavue, Nanterre, France
e-mail: adrien.krauz@gmail.com

D. Loorbach et al. (eds.), *Governance of Urban Sustainability Transitions*, Theory and Practice of Urban Sustainability Transitions,
DOI 10.1007/978-4-431-55426-4_8

process coordinated by state actors in view of implementing sustainable development.[1]

Others are claiming or formulating the hypothesis of a bottom-up transition, in which 'transition' stands as a banner for the convergence of scattered and diverse, already present, social movements or social innovation initiatives (including localisation, commons, voluntary simplicity, collective capabilities, transition towns) (Rumpala 2013; ATTAC 2013).

This dual opposition between an institution-driven transition promoting technological innovations or top-down incentives towards behavioural change, and a more grassroots concept, leaning on social innovation, communities, and civil society, is exacerbated by the French political culture. The French republican ideal was constituted through the ousting of all intermediates between citizens and state (i.e., communities, corporations, clergy), resulting in a gap between citizens and a centralised state, and leading to a framework where the bargaining power is a condition to societal change. The rise of the 'participatory and deliberative imperative' in public policies is symptomatic of this gap, questioning the technocratic legitimacy in defining common interests, opposing expertise and local knowledge, and questioning the elected representatives' monopoly to make decisions (Sintomer and Blondiaux 2002).

Nevertheless, these two approaches to transition do not exclude each other, even if the possibility of a 'bottom-up' approach challenges the capacity of institutional agents to implement transition in more than only a discursive way. In fact, several discourses around 'ecological transition,' developed by non-grassroots actors, lean towards innovations that are born in a grassroots context (e.g., local currencies, sharing economy) (Frémeaux et al. 2013; Grandjean et al. 2014).

Transition management (TM) is a governance framework based on a collaborative arrangement (the transition arena), aimed at involving relevant actors from different backgrounds in the envisioning of more sustainable futures and in an engagement with innovations and experimentations (Kemp and Loorbach 2006). These actors can be active in niches or regimes, and carry moderate, radical innovative, or transformative power (Avelino and Rotmans 2009).

In a TM context, how can top-down and bottom-up transition approaches engage in a dialogue? What forms of hybridisation can be produced, when regime or institution-based agents and niche-based or more grassroots actors meet in a transition arena?

8.1.2 Which Definition of Sustainability?

Sustainability is a vague concept that encompasses a range of discourses and political ideologies. It carries a varying transformative dimension, from 'status

[1] See the National Debate on Energy Transition, the creation of the National Council for Ecological Transition and the National Strategy for an Ecological Transition Toward a Sustainable Development steered by the Ministry of Ecology (MEDDE 2014).

quo' or 'ecological modernisation' where change is implemented by managerial, incremental, and top-down processes, to more transformative approaches where change is the result of political action, within and outside institutions (Hopwood et al. 2005). These transformative approaches are based on cultural claims, on the destruction of everyday life, and the suppression of the existential autonomy of individuals and communities (Gorz 2008). The result is tension between the scientific ecology, an 'expertocracy,' based on the definition of thresholds delimiting the carrying capacities for ecosystems, decided by experts and institutions, and the self-limiting 'political ecology,' defending everyday life culture, autonomy, and individual and collective self-capacities. This tension can be the cause of 'incommensurability,'[2] which can happen between discourses that hold no common foundations and give rise to potential conflicts and divergence between the two strands of transition as just defined.

In transition studies, sustainability is not defined as a substantive concept (Paredis 2013) but more as a discursive playing field (Robinson 2004) and "an open-ended orientation for change allowing pluralistic approaches in a deeply political and participatory process" (Grin et al. 2010, p. 2). These definitions of sustainability may allow the emergence of negotiated visions just as it can enable dispute over the multiple values and worldviews embedded in it. The focus of transition studies on technological innovation seems to shift towards an approach based on social and grassroots innovation (Seyfang and Smith 2007), empowerment, and social learning (Avelino 2011; Schäpke et al. 2013). This approach contributes to giving an increasing importance to an everyday life, grassroots sustainability, and calls for the potential negotiations, convergences, or hybridisations with more institutional or modernising versions of sustainability.

How can the implementation of TM manage the tensions between change dynamics which carry opposed conceptions of sustainability? Can these opposed positions be challenged, and what kind of definition of sustainability can emerge from a process that involves dialogue and negotiations between actors?

8.1.3 Transition Management in a French Context

TM is quite new in France and the academic and policy literature on sustainability transitions is hence almost nonexistent. The few authors interested in TM question the possibility of its implementation outside of the Netherlands, where "the capacity to reach consensus and to mobilise lifeblood around environmental issue is quite unmatched in the world" (Boulanger 2008, p. 73).

This chapter draws on the case of the city of Montreuil, the third most populous suburb of Paris, which was involved in a transition management process from 2011

[2] 'Incommensurability' is a translation of 'incommensurabilité', which is a neologism forged by Chateauraynaud (2013).

to 2014 as part of the elaboration of an environmental policy plan (the Local Climate Plan).

The chapter is based on a 2-year participant's observation from within the project team; the author has been part of the team as he was working for the Local Energy Agency, which contracted with the municipality to help implement the process. The uncertain and innovative nature of the project led the author to be given various tasks (from interviews and desk research to design and facilitation of the process) that progressively modified his position from a participants' observer to a more observant participator. This position can be defined as an "intellectually committed participation," where "the intensity of the participation darkens for a moment the lucidity of the researcher and its intellectual receptiveness" (Soulé 2007, p. 134). At times, the participation in project activities became a full-time job, often limiting the opportunity of remaining an observer and permanently maintaining the distance needed towards the research object. Thus, the methodology used for this article was mainly inductive research, relying on field notes and informal interviews.

Moreover, the radical novelty of TM and the field of sustainability transitions in the French context involved the adoption of a critical point of view towards TM as a research object itself. Thus, this chapter has a hybrid dimension: it questions the adaptation of TM in France as a new and exogenous methodology for accelerating societal change and at the same times also describes the implementation of TM from a practitioner's point of view.

In a national context of strong centralisation of power by the state and institutional actors and a politicisation of environmental issues in civil society, how can persons with different status (e.g., representatives, citizens, civil servants, activists) cooperate, and on which terms. Is the 'hybridisation' of perspectives, roles, and cultures, possible within a TM process?

8.2 Context

8.2.1 Historical Context

The story of Montreuil can be told as a succession of transitions. From the Middle Ages to the middle of the nineteenth century, Montreuil was a small agricultural suburb, its economy depending on the vineyards and the cultivation of peaches, which were grown in the shade of specific walls. With the Industrial Revolution, the settlement of numerous industrial activities entailed a massive urbanisation process, which progressively replaced the agricultural use of the land. In the 1960s, deindustrialisation caused massive unemployment. Forty years later, a new economic dynamic emerged with the expansion of the service sector.

Today, Montreuil is the fourth most populated city within the Parisian metropolis with more than 100,000 inhabitants. Montreuil is a complex city with a pivotal situation within the metropolis, torn between Paris, the 'Mother city' (Paris is so close that sometimes Montreuil is called "the 21st district of Paris"), the richer southern suburbs, and the more deprived northeastern suburbs. Thus, the city is cut in half: a topographic frontier (materialised by a strong slope and a highway) reinforced by a socioeconomic factor separates the Heights of Montreuil with little access to public transportation and predominantly social housing, from the Lower Montreuil, which is close to Paris and more attractive as a place of residence.

Its population, formerly moulded by successive waves of immigration, is very diverse but also contrasting: the arrival of a richer population during the 2000s resulted in the gentrification of the Lower Montreuil. After 70 years of municipal communism, a Green mayor and former Secretary of State for Ecology was elected in 2008, thus turning Montreuil into the most important Green city in France. The last municipal elections in 2014 led a coalition of ecologists, socialists, and communists to the City Hall with a new communist mayor.

Within this framework, sustainable urban development has to contend with climate change mitigation and energy transition, but also with the reduction of social inequities, insalubrious housing, fuel poverty, unequal access to public transportation and employment, etc., in a national context, where cities have very limited room to manoeuvre within the state or regional policies.

8.2.2 *Case Context*

Following the election of the Green municipality, the city administration made a commitment to climate change mitigation by ratifying the Covenant of Mayors in 2009, thus engaging itself to go beyond the '3×20' EU goals (by 2020: attain 20 % of renewable energy production; reduce CO_2 emissions by 20 %; increase energetic efficiency by 20 %) and to achieve Factor 4 by 2050. To reach these ambitious goals, the elaboration of the Local Climate Plan (PCET) was launched simultaneously with the Local Agenda 21.

The Local Climate Plan acts on an internal scale (by reducing all CO_2 emissions directly linked to City Hall activities such as buildings, car fleet, public lighting, and the increasing of municipal renewable energy production); a territorial scale (managing CO_2 emissions sources on which the City directly impacts through its jurisdiction, for example, housing, public transportation, urbanism), and an external scale (by mobilising local stakeholders in climate change mitigation).

The city applied for the 5-year EU-funded program MUSIC (Mitigation in Urban Areas, Solutions for Innovative Cities) mainly for the subsidies offered to support pilot projects, such as, in this case, an energy-producing school building. Transition management (TM) was embedded in the external part of the PCET, for its potential in the experimentation of new ways in mobilizing local stakeholders through the innovative approach of citizen participation.

8.3 Process Description

8.3.1 Creating the Transition Team (January–April 2011)

The first step of the TM process was the creation of a transition team, through which all the activities related to the project were to take place. It was implemented by a transition expert from January to April 2011 and accompanied by intensive team-building workshops, acculturation to TM's systemic perspective and concepts, and framing of the first insights concerning the city's transition analysis.

The transition team consisted of a five- to nine-member staff mobilised over a period of 3 years. Although the TM process was embedded in the Local Climate Plan and led by the officer in charge from the Environment Department, the members came from different City Hall departments (such as the Environment, Economic Development, Urban Development, Head Office, Social Urban Development) and from the Local Energy Agency, an NGO funded by local and national institutions to help implement policies related to climate change and involved since the very beginning of the project.

The specificity of this transition team resides in the horizontality maintained between its members and the diversity of its members' profiles. Although the project was under the responsibility of the Department of the Environment, the civil servants involved in the team enjoyed a relative autonomy vis-à-vis their respective departments. The transition team constituted a very unique 'protected space' where its members could develop a new cross-cutting understanding of the city, while experimenting with new working routines, without being tied by their accountability or obligation of results. Furthermore, they worked far beyond their assignment.

8.3.2 Gathering Frontrunners in a Transition Arena (April–September 2011)

The transition team consolidated itself with the recruitment of 'frontrunners' who are interested in joining the transition arena. These 'frontrunners' are change agents who are likely to bring their expertise, knowledge, influence, or creativity to the transition process. They were chosen for their open-mindedness, curiosity, listening, their attachment to the city and vision of its future, their ability to propose creative solutions, and their recognised expertise. Almost 40 change agents were identified by snowball effect, and their interviews were held between April and September 2011. The transition team put a great deal of effort into reaching gender balance and recruiting a wide range of actors with different social and professional backgrounds, living in both the Heights and the Lower Montreuil, to counteract the effects of the sociospatial structure of the city. This effort was made to open the process beyond the 'usual suspects' such as local sustainability experts or NGOs already involved in the Climate Plan or the Agenda 21, as well as to include 'difficult-to-reach' actors coming from the corporate world or from active migrant communities. In spite of these efforts, they unfortunately stayed out of reach: the corporate people approached by the transition team were disconnected from the social dimension of sustainability in Montreuil, and those from the migrant communities were trying to sustain their livelihood first and foremost and were thus far from ecological sustainability concerns. As a result, the majority of the transition arena consisted of white middle-aged male experts, who were at least self-aware of their non-representativeness of Montreuil's population as a whole. The profile of the arena was mixed with frontrunners active in the field of sustainability (architects, engineers, designers, ecologists) or other social fields (alternative housing development, bike school, social integration, education, international cooperation, culture), all of which facilitated the emergence of broad perspectives concerning the sustainability issues in Montreuil.

8.3.3 *Transition Arenas (September 2011–March 2013)*

The interviews, alongside relevant policy documents, served as the main information source for the transition team to produce a transition analysis highlighting the core values of Montreuil (diversity, proximity, experimentation, quality of life, citizenship), its 'niche city' character, where innovation and alternative cultures and practices are the norm, and the identification of six sustainable resources (economy, housing and buildings, urban nature and agriculture, mobility, conviviality and participation, social cohesion). This analysis was presented to approximately 25 of the interviewees during the first arena session (Arena 0) in September 2011, where they reacted positively to the work done by the team, while stressing the necessity to find new cooperative relationships between civil society and the municipality, and to bring together strategic thinking and action. At this point, the process was framed as challenging traditional participation by experimenting co-creation between the municipality and frontrunners.

Arena 1 in November 2011 saw several frontrunners voice their criticism. They found the approach too theoretical: the focus on sustainable resources and their evolution, the use of the TM vocabulary, and heuristic tools such as 'regime' and 'niche'—which, if not translated, appeared as jargon—and the S-shaped transition curb were too abstract in their opinion. They needed to appropriate both the methodology and the content of the transition analysis. Arena 2 in January 2012 was designed to overcome these critiques and to allow the participants to reframe the analysis, rename the resources, and identify different or new challenges for the city.

This step resulted in a broadened perspective on sustainability. From the original focus on climate change mitigation and energy transition, the new perspective emphasised the social, cultural, and political dimensions of ecological sustainability and the role of citizens and their tacit or traditional knowledge in inventing sustainable lifestyles. The outcome of these first sessions were recorded in a booklet and handed out to the participants.

After this first phase of analysis, Arena 3 (March 2012) consisted of envisioning a desirable future for Montreuil in 2030. A vision was built around three main ideas: a low-consumption (or modest) city, an emphasis on solidarity and social cohesion, and the consolidation of local democracy. The vision was not elaborated in a narrative form: it was in fact deepening the appropriation of the process by the frontrunners, extending the discussions around their vision of sustainability, including the emergence of underlying values. Thus, the transition was conceived as a global process, depending on existing bottom-up initiatives and a reform of local democracy, based on emergence and not on control, as well as creating social cohesion.

Arenas 4, 6, and 7 were dedicated to the elaboration of innovative projects that could lead or fit into the vision of the future or the guiding principles. During the first session (June 2012), ideas were collected to answer the four challenges framed as 'how to' questions (e.g.: 'How to turn Montreuil into a 'front running' city in terms of sustainability?'). More than 300 ideas were harvested, classified in a matrix ('Is this idea little or very innovative? Is this idea easily achievable or not?'), and selected by a vote. The second and third sessions (November 2012 and January 2013) were dedicated to participants working on 'solution cards,' which depicted draft projects solving the challenges through creative techniques of idea clustering and individual and collective narration. Seven 'solution cards' were thereby obtained, ranging from utopian (mobile city) to more concrete (a transition school) projects, illustrated by a short story of the solution, a description, and a slogan; these were to be connected with existing initiatives.

In the meantime, the whole process raised several uncertainties: the participants did not know what to expect from their participation in the arena; they were suspicious of the outcomes of the TM process; they contested being called 'frontrunners'; they questioned their legitimacy to work so closely with the municipality; and wondered what the concrete outcome of the process would be. This hesitancy entailed a reflection on the status of the arena, on its permanence, its continuity, and on the promise that their participation and dedication to the project would be effective. Thus, Arena 5 (September 2012) was designed as a regulation session, where the group could reflect on its purpose, its role, and its governance. The frontrunners and the transition team, in the presence of the elected representatives, agreed on a document determining the conditions for further cooperation.

Arena 8 took place in March 2013 in the City Hall Council Room. As the last session before the official launch event of the transition agenda, time was set aside for reflecting upon the whole process and to think about the future of the group, the role of each participant in the dissemination of the results, and the implementation of the projects.

The transition agenda was presented and symbolically ratified by the mayor during a launch event held in June 2013, where the *FabLab*, a project of an arena member, was jointly inaugurated.

8.4 Outcome

At first, the municipality had no exact idea what to expect from its involvement in the TM process. The elected representatives that were in charge of the project gave carte blanche to the transition team. Transition management was embedded in the external part of the Climate Plan to ensure its animation and promotion in a way to reach and mobilise local actors in climate change mitigation. Thus its main focus was to renew regular or traditional participatory approaches, which had proved difficult to become more than information or consultation tools (Arnstein 1969), by experimenting co-creation of public policies. Indeed, the participants' interviews showed that some of them were dissatisfied with the participatory phase of the Climate Plan, whose form predefined the outcome of its process.

Progressively, the project's stakes shifted as the adaptation of TM to the local context resulted in going from 'co-creating public policies' to 'how to create the conditions for such a co-construction.' Beside the co-construction of the transition agenda, a process involving reflexive thinking for the transition team was set, aiming at facing the uncertainty of the project and the role of civil servants in the transition.

This reflexive process allowed the civil servants to gain a better understanding of the uncertainty that characterises innovation and thus constitutes a real change vis-à-vis their usual practice. They learned to reduce this uncertainty by optimizing the team's 'self-management' and recognizing its benefits, because a unique 'protected space' has been created for this process. It allowed for a full appropriation of the process by the team members who participated in creative workshops aiming at identifying the desired outcomes of the project and at envisioning its effects in the long term. A new shared understanding of sustainability, reinforced by the inspiration provided by this creative work, was another outcome of this process parallel to an understanding of the responsibility to give impulse to change in an institution by developing a 'front-running' position within the organisation.

The transition agenda was a synthetic document summarizing the work done during the arena sessions. Attention was paid to make it didactic by trying to explain the whole process and its stages so that the method could be reproduced or serve as an inspiration for others. It consisted of the transition analysis along with the transition challenges, guiding sustainability principles, the vision for 2030, and then the seven 'solution cards,' plus some recommendations concerning the conditions for the transition to be achieved (Table 8.1).

After the launch event in June 2013, the process did not end.

The team knew that the municipality did not possess the funding necessary to single-handedly implement the projects envisioned during the arena process. Moreover, municipal elections were coming up in 2014, marking the beginning of the campaign season and thus an unfavourable period for the development of new or innovative projects. The team decided to use the participatory dynamic, which was initiated by the transition arena, to complete a roadmap for the implementation of the draft projects. They planned a 2-day festival (the 'MUSIC and FabLab

Table 8.1 An overview of the main elements of the transition agenda

Sustainable sector	Transition challenge
Economy	How to relocate the food supply chain and make economic activities become more just and local?
Population	How to make the energy transition a shared and inclusive project of society as a whole?
Building and housing	How to structure the environmental building sector?
Nature and urban agriculture	How to develop new relationship with the environment that is based on partnerships and cooperation?
Social cohesion	How to put social cohesion and participation at the core of every project?
Mobility	How to root the bike culture in the spirits of users and designers?

Guiding sustainability principles
Energy transition is not an isolated transition. It implies a political, social and cultural transition
Transition rests on existing innovative practices and experiments
Social cohesion as a guiding principle
Public space is the medium of the transition
Relocalization as a 'new horizon'

Shared vision
'A city where energy is produced and is used in moderation, shared fairly while creating social cohesion, relying on a new relationship between the municipality and the citizens'

Solution cards	
1. Mobile city	A new way of living perpetual change
2. Sharing space and knowledge	Bringing together unemployed people and ecological construction companies to share knowledge and build co-housings
3. City metabolization	Alternative waste and water management using natural metabolization processes
4. Productive rooftops	Using rooftops to grow food, harvest water and produce energy
5. Participatory building retrofit	Facilitating do-it-yourself ecological building retrofit
6. Creating 'energy cohesion' for 'social renovation'	Promoting ecological retrofit projects with festive or artistic events
7. Transition school	A perennial festival associated with the *FabLab* where learning, sharing, gathering, making, producing and experimenting is possible

Festival') alongside with frontrunners to disseminate the transition agenda and the 'solution cards' produced by the arena. The first day was dedicated to eco-design, with demonstrations of the *FabLab* and workshops. The second day, the team gathered a panel of project holders, all connected to or inspired for the future implementation of the draft projects. A hundred people interested in transition experiments and eco-design could take ownership of the projects by attending workshops facilitated by transition team members, who applied the same creative techniques used during the arena sessions.

This dynamic of dissemination was difficult to maintain because of the election period and the perspective of a political change, and hence projects were mainly taken up by actors from the transition arena. The Local Energy Agency handled the projects related to energy-efficient building renovation, a cooperation between a frontrunner and a social housing organisation was set up for the use of rooftops to produce food and energy, and the *FabLab* was fueled by the dynamic of the transition arena. Still, the process fostered new cooperation between different stakeholders despite the lack of funding or human resources within the municipality.

Besides the projects or transition experiments envisioned by the transition agenda, the second main activity after the launch of the transition agenda was dedicated to the prefiguration of the Local Transition Council (CLT), identified as a key element for the creation of favourable conditions for accelerating the transition. The CLT serves as a step towards an institutionalisation of the transition arena, which appears necessary regarding the political context to reinforce its democratic legitimacy and regulate its interactions with 'normal' policy. Two additional arena sessions took place to reflect upon its composition, functioning, funding, role and governance. The CLT was defined as an open council, whose role is to support transition experiments, the incubation of projects and ideas, but also to raise public awareness regarding transition-related issues. The creation of this council is now the object of a political negotiation among the members of the coalition, who won the elections in March 2014. Nevertheless, the context is sensitive; the CLT has been politicised during the election campaign by members of the arena, who created an 'election list' that did not aim at being a candidate but only proposing ideas, resulting in anchoring this proposition as a "Green" one, and making it difficult to appropriate for the new communist municipality. The former "Green" majority that was aware of the process is now a minority member of the coalition which won the elections. The CLT, if voted, will be dependent for its funding on the municipality.

8.5 Discussion: Hybridisation, Evolution of Roles, Relations, and Cultures within the Project

In this section, we reframe the process from the perspective of hybridisation between top-down and bottom-up logics, between institutional actors and citizens, leaning towards a redefinition of traditional roles and an inversion of power relations, resulting in a new shared 'transition culture.'

The TM process in Montreuil showed a specific arrangement involving institutional and civil society actors. The composition of the transition team was already a hybridisation between civil servants and the energy agency, which holds an ambiguous semi-institutional position (it is an NGO created and funded by public institutions). This arrangement entailed different strategies and interests from each actor.

The energy agency has a fine-tuned knowledge of climate and energy issues and actors on the ground, but it has a fragile structure dependent on institutions and thus it is in search of legitimacy and autonomy.

The city contracted the energy agency for its expertise, but the construction of the project entailed an ambiguous relationship between these two organisations: both symmetrical (horizontality within the transition team) and asymmetrical (the city was the contractor). This lack of role definition allowed for the evolution of this relationship during the process.

The elected representatives were fully supportive of the TM project and gave carte blanche to the transition team to implement it. Nevertheless, the free hand served as a mask to conceal a lack of defined goals or expectations assigned to the TM process as a whole. This lack allowed for uncertainties to rise, which were progressively taken up by the energy agency, thus empowering it and allowing it to have a more leading role, reframing the stakes of the process, and leading a shadow track process within the transition team itself.

8.5.1 The Role of the Energy Agency: Instigating Reflexivity

The leadership of the TM process progressively shifted from the team of civil servants to the local energy agency. This change came as a direct result of the reformulation of the goals originally defined in the tender, as well as of a reflection on the adaptation of the process to the local context.

Starting as a way to experiment new ways of mobilising citizens concerning climate change mitigation, the process was reframed around three goals. First, it could contribute to the creation of space for cooperation between the municipality and citizens or civil society. As seen in the introduction of this chapter, the rise of a participatory and deliberative imperative in French public policies questions the legitimacy of experts in the definition of common interests and the elected representatives' monopoly to make decisions. The transition arena, with its emphasis on collaboration, negotiation, and dialogue between relevant actors to engage in innovation, was perceived as an opportunity to set up a 'protected space' where new cooperative relationships between citizens, civil servants, and officials were to be experimented. The difference with the original 'mobilising citizens' framework lies in the idea of a more horizontal and open process.

Second, this cooperative space could allow for the reframing of sustainability issues from the citizens' standpoint, aiming to go beyond institutional or expert versions of sustainability solely focussed on climate change mitigation. This could in return allow for the exploration of other conceptualisations of sustainability grounded into the daily practices and cultures of the diverse inhabitants of Montreuil.

Last, the process could encourage institutional reflexivity (defined as the capacity to engage in 'double-loop' learning processes and change in governing values (Argyris and Schön 1974)), as a prerequisite to co-create with citizens and civil society.

This last point led the energy agency to create a shadow process within the team, identifying the lack of co-creation culture amongst civil servants as a persistent problem. This shadow process aimed at making the team members take into account the reflexive dimension of the project and answer the following underlying question: 'What is my role as a civil servant, working in and representing an institution, in this transition?'

Thus the energy agency took the role of an instigator of reflexivity (Demoulin and Tribout 2014), protecting the process against the most powerful player involved (the municipality).

8.5.2 *Exchanging Traditional Expertise Roles: Civil Servants as Receivers of Citizens' Voice, Frontrunners as Inspirators of Future Public Policies*

During the recruitment phase, which consisted in interviewing potential frontrunners, members of the team were trying to identify individuals matching the profile for which they were looking. This activity took place both at and away from work: they asked their colleagues, friends, and relatives whether they knew such individuals. They gained a better knowledge of the local actors involved with sustainability concerns, of the networks of change agents in the city, and they discovered inspiring people. The interview process was a mine of information that engaged people with a genuine point of view, a vision of the city, its persisting problems, and its future.

This phase gave members of the team, working in an institutional or 'regime' context, access to actors from the 'niches,' developing more or less radical discourses on sustainability, including critical views on the present municipal policies. All in all, the team enriched their understanding of the city and of its future prospects.

The first transition analysis followed from the interviews and its harvest of local knowledge; the team analysed all the material resulting from the 40 interviews to produce a synthesis presenting six 'sustainable resources' (economy, housing and buildings, urban nature and agriculture, mobility, conviviality and participation, social cohesion) and transition challenges for each of them.

The transfer from informal citizens' voice to a structured transition analysis entailed conscious and/or unconscious modifications, so that the content of the analysis could be compatible with the way the city framed its environmental policies (e.g., the main focus on CO_2 reduction or energy savings to fit the goals stipulated in the Covenants of Mayors).

To overcome this institutional reframing, an arena session was designed to discuss the definitions of transition concepts, challenges, and persistent problems, so that the first transition analysis could be deconstructed, then reconstructed from a frontrunner's standpoint, revealing new arrangements and new framings of the problems. The reframing of the challenges raised by the arena introduced a vision of sustainability different from those underlined by the institutional discourses, in particular with regard to the strong focus on energy and climate change mitigation.

By giving legitimacy to local or citizen's knowledge (Nez 2012), the transition management process inverted the expertise roles. The civil servants were led in a learning process to build on the inspirations and expertise provided by the frontrunners. Moreover, the formulation of the transition analysis was the result of a back-and-forth shuttling between these change agents and the civil servants, who were invited to leave their institutional and expertise positions aside in favour of drawing on their capacity to harvest and value local knowledge. This reframing of civil servants' mission as facilitators and animators of co-created public policies is the key to governing urban transitions and deserves special attention.

8.5.3 New Cooperative Relationship between Civil Society and Institutions

During the first arena session (Arena 0), the question 'How can we create new cooperative relationships between the municipality and the citizens?' was raised by both the team and the frontrunners. The city councillors in charge of the project guaranteed that they would not interfere or take over the project and its results, thus reasserting the carte blanche given to the team for this project. One of the frontrunners even added that, on the contrary, it was necessary that elected representatives should exploit or use the ideas produced within the arena. An exercise, which consisted of asking the participants where they estimated their current practices to be on a scale ranging from 'institutional' to 'independent' and where they see themselves in the near future on the same scale, showed clearly their willingness to bring actors together from both institutions and from civil society.

The relationship between the team and the frontrunners evolved during the next arena meetings. The team was subjected to criticism from the frontrunners (concerning the municipality's management or projects) before mutual trust could be established.

The team had a more and more active role into the facilitation of the arena sessions and progressively took active part in the activities. The status of the members of the team changed during Arena 5, when they drafted a charter of mutual commitment between two stakeholders: the city, composed of the transition team, a local councillor, and the energy agency, and the transition arena, composed of the frontrunners. The frontrunners further added that the transition arena included the transition team, because of its involvement and its active contribution to the elaboration of the outcome of the process. The charter comprised governance rules for the transition arena, such as respecting horizontality in decision making, collegiality, mutual respect, consensus, zero criticism principle, and the common ownership of the ideas elaborated within the arena.

The emergence of this cooperative space allowed the group to go beyond the traditional hierarchical and often confrontational relationships between representatives and citizens and to fill the gap that exists between citizens and institutional actors, thus bringing symmetry and horizontality, which are most unusual features amongst local participatory processes. Yet, the institutionalisation of such 'hybrid' space, which allows for incorporating 'reflexive moments' into policy making (Hendriks and Grin 2007), raises issues concerning the selectivity of the process, the legitimacy of its participants, and its embedding in a representative democracy framework.

8.5.4 The Emergence of a Shared Transition Culture

The transition arena, bringing together actors from civil society and institutions in a horizontal and cooperative way, inverting the traditional expertise roles, proved itself to be a space of negotiation for visions of sustainability. As we have stated, sustainability encompasses diverse discourses that have different postures regarding what unsustainability entails. The TM process, embedded in the Local Climate Plan, focused on CO_2 and energy consumption reduction, under an 'energy transition' label.

The initial challenges raised in this transition were displaced by the introduction of the social dimension of sustainability, stressing that energy transition is not an isolated transition; it implies a social, cultural, political transition, which became in retrospect the first guiding principle formulated by the arena. Montreuil's arena stated that the transition towards sustainability is more than a matter of politics; it entails radical lifestyle change and a social and cultural transition. Solutions proposed by the arena evoked grassroots conceptions of sustainability: simplicity as a way of life, conviviality, and low-tech solutions, all partaking in reinforcing social cohesion within the very diverse population of Montreuil. The dimension of transition as a change process was underplayed to the benefit of transition as an empty signifier (Laclau and Mouffe 2001), articulating a chain of equivalences (social bound, equity, conviviality, inclusiveness, sobriety, ingenuity, etc.), and it participated in politicizing sustainability by reframing it as an alternative bottom-up vision of society.

8.6 Conclusion

This chapter aimed to describe the adaptation of TM within a French metropolitan context, where building coalitions of stakeholders or finding consensus is not part of the political culture because of the prominent role of the state and public institutions. The discussion analysed the possibility of creating hybridisation patterns between institutional players and actors from civil society: between their status, their sustainability visions, and their practices.

Involving the energy agency in the process and adopting the position of instigating reflexivity amongst civil servants allowed for the formulating of the need to adapt TM methodology to the political context of the city and to the French political culture in general.

This position aimed to go beyond the institutional framings of sustainability, mostly involving top-down incentives towards behavioural change and technological innovation to mitigate climate change and to bring about more grassroots and bottom-up ideas of sustainability. These ideas were closer to political ecology: that is, promoting self-limitation, defending everyday life culture, autonomy, and individual and collective self-capacities. It helped focus on conserving the radical dimension of the analysis proposed by the participants and to avoid any institutional reframing.

This adaptation caused the procedural dimension of TM and its practical aspects (envisioning, backcasting, elaborating and choosing transition paths, monitoring, etc.) to be underplayed at the expense of its substantive dimension (the transition), as we saw the focus on the energy transition or post-carbon transition vanish in favour of a broader analysis leading to a transition equivalent to a system change (cultural, political, and social transition). This analysis was widely shared by both civil servants and frontrunners, who all contributed to the creation of a transition culture, nurtured by the ideas brought on by the participants. The transition arena thus became a channel for new or more radical ideas to filter into institutional structures.

Another aspect of this adaptation is the emergence of the need for a transition of institutions, raised near the end of the process by the Head Executive of the Department of the Environment of the city. Transition management states that current institutions are inept to handle persistent or wicked problems caused by modern problem solving. Reflexive governance can be a framework in which more sustainable decision making can be made, in particular regarding the development of strategies and goal formulation in interaction with a wide range of actors. The role of civil servants in this transition is crucial; they are situated at the interface of elected representatives, enjoying strong legitimacy and an expert status, and of citizens, who have no democratic legitimacy but have knowledge from their daily life in the city, from their activist engagement or their professional activity. They can help make "innovative conceptions of living generated in civil society [...] unsettle established practices and challenge the state" (Meadowcroft 2011, p. 73). Their empowerment, through the learning of new skills (e.g., facilitation, animation, systemic thinking, uncertainty management) and the confronting of their institutional or expert knowledge and routines with that of citizens, is key to transforming public action into a more sustainable and a more democratic way.

References

Argyris C, Schön DA (1974) Theory in practice: increasing professional effectiveness. Jossey-Bass, San Francisco

Arnstein S (1969) A ladder of citizen participation. J Am Inst Plann 35(4):216–224

ATTAC (2013) Petit manuel de la transition: pour toutes celles et ceux qui aimeraient mais doutent qu'un autre monde soit possible. les Liens qui libèrent, Paris

Avelino F (2011) Power in transitions. Empowering discourses on sustainability transitions. PhD dissertation. Rotterdam University

Avelino F, Rotmans J (2009) Power in transition: an interdisciplinary framework to study power in relation to structural change. Eur J Soc Theory 12:543–569. doi:10.1177/1368431009349830

Boulanger PM (2008) Une gouvernance du changement sociétal: le transition management. Rev Nouv 11:61–73

Chateauraynaud F (2013) La radicalité est-elle soluble dans l'argumentation ? La sociologie des controverses et l'endogénéisation de la critique sociale. Paper presented at the doctoral seminar Fructis Arc, Liege University, 17 Apr 2013

Demoulin J, Tribout S (2014) Construire des espaces de réflexivité pour analyser et transformer les pratiques professionnelles: un travail de légitimation, ¿Interrogations? 19

Frémeaux P, Lalucq A, Kalinowski W (2013) Transition écologique, mode d'emploi. Les Petits Matins, Paris

Gorz A (2008) Écologica. Galilée, Paris

Grandjean A, Le Teno H, Hulot N (2014) Miser (vraiment) sur la transition écologique. Les Ed. de l'Atelier, Paris

Grin J, Rotmans J, Schot J et al (2010) Transitions to sustainable development: new directions in the study of long term transformative change. Routledge, New York

Hendriks CM, Grin J (2007) Contextualizing reflexive governance: the politics of Dutch transitions to sustainability. J Environ Policy Amp Plan 9:333–350. doi:10.1080/15239080701622790

Hopwood B, Mellor M, O'Brien G (2005) Sustainable development: mapping different approaches. Sustain Dev 13:38–52. doi:10.1002/sd.244

Kemp R, Loorbach D (2006) Transition management: a reflexive governance approach. In: Voss JP, Bauknecht D, Kemp R (eds) Reflexive governance for sustainable development. Edward Elgar, Cheltenham

Laclau E, Mouffe C (2001) Hegemony and socialist strategy: towards a radical democratic politics. Royaume-Uni, London

Meadowcroft J (2011) Engaging with the politics of sustainability transitions. Environ Innov Soc Transit 1:70–75. doi:10.1016/j.eist.2011.02.003

Nez H (2012) Nature et légitimités des savoirs citoyens dans l'urbanisme participatif. Une enquête ethnographique à Paris. Sociologie 2:387–404. doi:10.3917/socio.024.0387

Paredis E (2013) A winding road. Transition management, policy change and the search for sustainable development. PhD dissertation. Universiteit Gent

Robinson J (2004) Squaring the circle? Some thoughts on the idea of sustainable development. Ecol Econ 48:369–384

Rumpala Y (2013) Sur les possiblités d'une transition écologique "par le bas": vers une théorie post-gouvernementale. Paper presented at the workshop "Développement durable et 'acceptabilité sociale'". Paris Val de Seine Architecture School, 15 Feb 2013

Schäpke N, Omann I, Mock MM et al (2013) Supporting sustainability transitions by enhancing the human dimension via empowerment, social learning and social capital. Proceedings of the SCORAI Europe workshops, Rotterdam, 7–8 Oct 2013

Seyfang G, Smith A (2007) Grassroots innovations for sustainable development: towards a new research and policy agenda. Environ Polit 16:584–603. doi:10.1080/09644010701419121

Sintomer Y, Blondiaux L (2002) L'impératif délibératif. Politix 15:17–35. doi:10.3406/polix.2002.1205

Soulé B (2007) Observation participante ou participation observante ? Usages et justifications de la notion de participation observante en sciences sociales. Rech Qual 27:127–140

Part III
Synthesis and Reflections

In Part III the focus is on synthesising, reflecting, formulating insights, and drawing lessons.

This part comprises three chapters, with the first one synthesising insights and lessons from the experiences with the (operational and heuristic application of the) transition management framework in the five European and Asian cities (Chap. 9). The remaining two chapters draw recommendations for practitioners about the governance of urban sustainability transitions (Chap. 10) and outline a research agenda for this topic (Chap. 11).

Chapter 9
Insights and Lessons for the Governance of Urban Sustainability Transitions

Julia M. Wittmayer

Abstract This chapter synthesises insights and lessons from the experiences with the operational and heuristic application of the transition management framework in five European and Asian cities: Aberdeen, Ghent, Higashiomi, Kitakyushu, and Montreuil. Zooming in on one of the transition governance principles, this chapter analyses how far the transition management activities outlined in the five empirical chapters of this book have contributed to creating space for change to building up alternative regimes and challenging the status quo. This analysis allows drawing lessons for transition governance principles and for operational process transition management designs.

Keywords Governance • Local context • Practices • Governance principles • Social relations • Transition arena • Transition management • Sustainability

9.1 Introduction

In this book, it has been argued that we live in a complex and uncertain world facing persistent problems that are deeply embedded in societal structures and multi-actor contexts. To address these problems, new forms of governance are needed such as transition management, which is increasingly applied in the urban context (Loorbach and Shiroyama 2016, Chap. 1, this volume; Wittmayer and Loorbach, Chap. 2, this volume; Nevens et al. 2013; Nevens and Roorda 2014; Wittmayer et al. 2015). Transition management suggests that emerging transitions can be facilitated, influenced, guided, and accelerated through transition governance interventions (Loorbach 2010; Loorbach et al. 2015; Frantzeskaki et al. 2012).

The preceding five chapters were focussed on either applying transition management in an operational way, in terms of a framework guiding the implementation of participatory processes (Hölscher et al. 2016, Chap. 6, this volume; Frantzeskaki and Tefrati 2016, Chap. 4, this volume; Krauz 2016, Chap. 8, this volume), or in a

J.M. Wittmayer (✉)
Dutch Research Institute for Transitions, Erasmus University Rotterdam, Rotterdam, The Netherlands
e-mail: wittmayer@drift.eur.nl

D. Loorbach et al. (eds.), *Governance of Urban Sustainability Transitions*, Theory and Practice of Urban Sustainability Transitions,
DOI 10.1007/978-4-431-55426-4_9

heuristic way employing it as an analytical lens to describe and analyse past and ongoing governance processes in cities (Mizuguchi et al. 2016, Chap. 5, this volume; Shiroyama and Kajiki 2016, Chap. 7, this volume).[1] This chapter analyses in how far the transition management activities as outlined in the five empirical chapters of this book have contributed to creating space for change to building up alternative regimes and challenging the status quo in the cities in question. As such, this chapter zooms in on one of the transition governance principles and its envisioned outcomes, namely, creating space for change to build up alternative regimes (as outlined in Loorbach 2007, 2010; Wittmayer and Loorbach, Chap. 2, this volume). Based on the systematic analysis of the five empirical chapters, I draw lessons for transition governance process design and principles

The chapter is structured as follows. First, I consider the local context and challenges that led to the uptake of transition governance activities (Sect. 9.2). Second, I analyse how transition governance activities led to the creation of spaces for interaction (Sect. 9.3), and third, which alternative ideas, practices, and social relations emerged in these spaces (Sect. 9.4). Fourth, lessons are drawn from the process design of operational applications of transition management and principles for transition governance (Sect. 9.5).

9.2 The Local Context and Challenges

The five cities (Aberdeen, Ghent, Higashiomi, Kitakyushu, Montreuil) are situated in different sociocultural, political, economic, and geographic contexts, all conducive to adapted versions of transition management (see Table 9.1 for an overview). The main difference is, of course, the European and Asian context, respectively. The cities range in size from just about 100,000 to nearly 1 million inhabitants, but share the overall challenge of addressing unsustainability as it presents itself in the different contexts. Comparing the five cities clearly shows the importance of the context for the ways that unsustainability manifests, whether it be super-ageing (as in Higashiomi), water and environmental pollution (in Kitakyushu), or oil dependency (in Aberdeen).

Existing structures, cultures, and practices also mediate what sustainability or sustainable development come to mean and which alternatives are developed. An example of an existing context is the historically laden relationship between citizenry, private sector, and public sector, which can pose challenges for multi-actor societal learning processes. Although in Europe the processes focused mainly on collaborations between citizens and public sector with a sporadic participation of

[1] While in three cities a specific prescriptive methodology was followed, the other two cities did not: navigating these differences has been a challenge in the analysis for this chapter. To address this challenge, I have been as transparent as possible in my reference to these chapters.

Table 9.1 Overview of local context, dynamics, and challenges in the five cities (adapted from Wittmayer et al. 2016, Chap. 3, this volume)

	Aberdeen	Ghent	Higashiomi	Kitakyushu	Montreuil
Context	210,400 inhabitants	240,000 inhabitants	120,000 inhabitants	970,000 inhabitants	103,000 inhabitants
	Coastal city in the North of Scotland	Harbor city in the East of Belgium	Located between Nagoya and Kyoto in the heart of Japan	Most northern city on the western island of Kyushum, Japan	In the immediate proximity of Paris, France
(Transition) challenges for urban development	Ambitious CO_2 reduction goals	Ambition to become climate neutral	Ambitious CO_2 reduction goals	Ambitious CO_2 reduction goals	Ambitious CO_2 reduction goals
	Create a culture of energy efficiency among public and private actors	Start of 'climate alliance', a network of diverse committed actors	Job creation in food, energy, care	Job creation in environmental fields	Create a culture of efficiency for a 'positive energy strategy'
			Super-ageing society	Super-ageing society	
Heuristic or operational application of TM	TM as operational application to support the development of alternative pathways towards urban sustainability and to mobilise collaborative arrangements between different city actors. TM also as heuristic to analyse the value of visioning	TM as operational application to explore what a climate-neutral future would mean for the city and to involve actors from the city into developing and implementing actions	TM as heuristic application to investigate the emergence of a multi-niche collaboration	TM as heuristic application to analyse the 'Kitakyushu Eco-town Project' as a catalyst for a historical transition from an industrial to a sustainable city	TM as operational application to mobilise local stakeholders in climate change mitigation, turned into a negotiation of meaning of sustainable urban development with a specific focus on governance

business, the Japanese context demonstrates a more pro-active role of business actors in promoting (especially sociotechnical) transition pathways. In all cases, the main focus of the public sector was (generally speaking) on the more tangible and ecological aspects of sustainability, whereas citizens highlighted and introduced social considerations in future pathways.

In the European context, developments that were having repercussions for the cities are ongoing demographic changes, the 2008 financial crisis and its impacts in terms of a changing welfare state and budget cuts, amongst others, stagnating employment opportunities, and high youth unemployment, combined with high CO_2-reduction targets. All three European cities—Aberdeen, Ghent, and Montreuil—have ambitions to tackle climate change by focussing on CO_2 mitigation. This goal can be traced back to their involvement in the MUSIC project, which aimed to develop climate mitigation measures and involve actors through a transition management approach (for more information, see Wittmayer et al. 2016, Chap. 3, this volume). In the transition management process, the search for a sustainable future also meant to open up existing narrow problem framings, to let go of the current identity of the city and to negotiate a new one. For Aberdeen, for example, this was a shift from envisioning the city as 'oil capital' to a 'low-carbon city.'

In the Japanese context, the Higashomi case clearly illustrates the threefold challenge that Japanese society is currently facing: the demographic challenge of super-ageing, the shift to sustainable energy, and the striving for healthy economic development. In Kitakyushu an ecological focus on environmental pollution and recycling prevailed, which corresponds to the Zeitgeist of this historical case. Over the course of the years, this focus opened up to include job creation and industrial development.

Logically, a major focus of all five cities is on the ecological aspects of sustainable development as they have been selected for inclusion in this book based on their affiliation with specific networks (Wittmayer et al. 2016, Chap. 3, this volume). However, in addressing these ecological aspects, all five cities opened up their problem framing to integrate other issues.

9.3 Transition Management as Creating Spaces for Interaction

One of the transition governance principles refers to the creation of space for change to build up alternative regimes (as outlined in Loorbach 2007, 2010; Wittmayer and Loorbach, Chap. 2, this volume). This principle has been further developed and transition management has been described as creating spaces for interaction between different kinds of actors that allow for alternative ideas, practices, and social relations to emerge (Wittmayer et al. 2014). These spaces are also described as allowing the process to be "*in the shadow of regular policy*" (Van Buuren and Loorbach 2009, p. 379).

Building on these ideas, this section identifies and analyzes these kinds of spaces, concentrating more on operational applications (focusing on creating such spaces), but including insights from heuristic applications (focusing on analysing existing governance processes in terms of such spaces) of transition management. The 'transition arena' and 'transition team' are analyzed respectively, as spaces for interaction, based on a distinction of different settings in transition management processes by Roorda et al. (2014). However, this should not gloss over the interaction between these two spaces: these are overlapping and constantly evolve alongside and with one another. The process in Kytakyushu City, for example, shows that there are overlaps in that members of the transition team also play a role in the transition arena, whereas in many of the European cities the transition arena was facilitated by a professional who belonged to neither the city administration nor the local community.

Transition Team

The transition team can be described as both an actor in the transition management process as well as a space for interaction. As an actor, the transition team performs a number of activities in participatory transition management processes, which have been described as follows: "*the team not only prepares, documents, analyses, monitors, co-ordinates, manages, facilitates and evaluates the whole process, but also chooses the participants and feeds them with background information and detailed knowledge. It brings together the various parties, is responsible for the internal and external communication, acts as intermediary in discordant situations and has an overview of all the activities in and between arena meetings*" (Wittmayer et al. 2011, p. 31). The transition team also constitutes a space in which different actors meet and collaborate to frame and facilitate a transition management intervention.

Dominant structures, cultures, and practices are questioned already in the transition team. The transition team consists of different kinds of actors coming together to collaborate and to prepare as well as to guide the participatory process. In the European cities, the transition team consisted of urban policy makers from different departments, transition researchers, and professional facilitators. These actors bring their different worldviews and perspectives as well as problem framings, which need to be continuously negotiated in the transition team to frame and implement the participatory process. The transition team members in the European cities were also learning while doing: based on their knowledge about transition management from a number of training sessions, they had to frame and implement an operational transition management process supported by transition researchers. The transition team in this respect was a space for learning, coaching, and experimenting.

The fact that policy makers from different departments are collaborating makes the transition team also a space for cross-departmental collaboration. In Montreuil, for example, the collaboration was described as non-hierarchical (i.e., decisions were taken collaboratively) and open (i.e., there was no pre-definition of SMART outcomes and impacts). According to Krauz (Chap. 8, this volume), this open collaboration allowed policy makers a new more integral understanding of their city and experimentation with new working routines without being tied to departmental obligations. In Aberdeen, the interaction in the transition team also challenged the current policy structure, which "*is not made for future-oriented thinking*" (Frantzeskaki and Tefrati 2016, Chap. 4, this volume). The transition teams in the European cities thus were crucial for intra-organisational learning as well as flagging the misfits of current local government structures, cultures, and practices with transition governance principles. This misfit can be traced to the status quo in local governments, which does not always encourage cross-departmental working, orientation on the long term (meaning one to two generations), or a focus on experimentation and learning.

The case studies also provide insights into the kind of roles needed in a transition team setting. As a motor of the overall process, frontrunners are needed not only as part of the transition arena but also for the transition team. The examples of Kitakyushu (where the team was composed of visionary business actors) and Ghent show that involving players who can '*play the regime game*' and have transformative power is crucial. These players can identify windows of opportunity in terms, for example, of using subsidies as seed money or identifying moments for strategically liaising with other actors. The Montreuil case shows that the involvement of a relative outsider, such as the local Energy Agency, in the transition team leads to a high degree of reflexivity of the team about their own roles and how they themselves can address inhibiting structures, cultures, and practices in their own environment.

Transition Arena

Similar to the transition team, the transition arena can be seen as both an actor in the transition process as well as a 'space for interaction,' in which different actors interact to address sustainability challenges and further a sustainability transition. Such spaces not simply 'are,' they come about in and through "*dialogical encounters between people*" (Wittmayer et al. 2014, p. 479). They also come about through both dialogue as well as action: one can question current understanding and formulate alternatives. Especially from the three European case studies, interesting insights arise in relation to the nature of the boundary of such a transition arena.

One way of describing the boundaries is by focusing on the aspect of time in its numerous respects. First, it refers to the question how long a transition management process may reasonably take. Although some actors feel rushed, others would not want to wait any longer until the vision is followed by actual experiments. In Ghent, the process took just about 1 year and the transition arena participants have felt pressured by the high pace and abrupt ending of the process. In Aberdeen, the process due to changes in the transition team took about 2.5 years. A hasty transition arena process might limit the ownership of participants over the process, but the process also needs to maintain a certain speed to sustain enthusiasm, create momentum, and seize opportunities. However, if compared to the time that an overall transition takes, an arena is nothing more than a temporary impulse (the historical transition of Kitakyushu took about 30 years). The second question is how much time one might reasonably expect actors to invest in this process. This point is especially challenging in the transition management processes organised by the European cities as many of the participants were participating on a voluntary basis, thus in their free time. Especially, the example of Ghent showed that those participants who could combine their work with their engagement in the transition arena process were more active in developing and implementing follow-up activities. For the Kitakyushu case, this question was less relevant as private and public actors were involved in their formal positions; it also was not relevant in Aberdeen, where participants attended meetings during their worktime. A last interesting 'time' aspect relates to the question about the 'right' time for a transition management intervention. Here, the example of Ghent shows that a targeted timing, that is, before municipal elections and as providing input for them, does improve uptake and impact. In Aberdeen, the transition management process came after a conflict between civil society and local government over the appropriation of the local urban gardens for commercial use and a lack of trust between various actors across sectors. In this context, the transition management process was used to kick-start a new trust-building process.

The boundary of a transition arena is also defined by the nature of the personal relations between transition arena participants and policy officers, respectively, as well as within each of these groups. The intense processes have led to closer relationships between transition arena participants in the cities. These intra-arena relationships, especially when based on trust, make the arena a group and thereby define an inside and outside, a kind of boundary. This boundary is fluent as it allows for people to join in and leave easily. In the case of Ghent, Hölscher et al., (Chap. 6, this volume) refer to political empowerment of participants and policy officers, which relates to the intra-arena relations, but also to relations with actors who joined after the transition agenda was formulated or the overlap between transition arena and transition team.

A last interesting insight relates to the fact that the framing of the role and status of the transition arena can be kept open deliberately, or that a closure of this framing can be implicitly or explicitly resisted. These actions can allow for a productive engagement with the plurality of possible framings and a collaborative, contextualised definition of role and status. This aspect has been specifically important in the transition arena process in Montreuil, where the initial lack of such a definition led to a collaborative, although not conflict-free, search. The status of the arena (i.e., its role, duration, competencies) and the legitimacy of the process and outcomes (i.e., relation to representational democratic processes, etc.) had not been carved in stone in the beginning; rather, the transition team also was learning about using transition management, which opened the possibility for an explicit and productive discussion about what such a role of the transition arena *could* be. In Montreuil, this turned out to be paving the road for the establishment of a Local Transition Council. In this regard, although a framing is needed it might be wise to resist an early and especially one-sided (driven only by the transition team) closure thereof to allow for productive uncertainty and openness. A necessary condition might be the 'carte blanche' that Montreuil's transition team received from its local council in relation to its transition management activities.

9.4 Transition Management as Challenging the Status Quo

In creating spaces for interaction such as the transition team or transition arena, the status quo in the five cities was challenged in a number of ways. It is in these spaces that "*alternative ideas (e.g., knowledge, discourses, visions), practices (e.g., transformative action, experimentation, learning) and social relations (e.g., actors) that further sustainability transitions are developed and nurtured*" (Wittmayer et al. 2014, p. 470; cf. Loorbach 2007, 2010). After offering some general observations, I scrutinise the alternative ideas, practices, and/or social relations that emerged in the different cities. These three outcome dimensions do also overlap and influence one another, and the separation here is made for the sake of analysis.

A first general observation is that every city seemed to have a specific focus on one of these three dimensions. In Higashiomi or Kitakyushu, the focus was on the establishment of alternative practices, whereas in other processes the focus was on more political questions, such as the framing of social roles and relations, as in Montreuil. In still other processes, the primacy was on the development of new ideas, as in Aberdeen. This does not mean that the other two aspects are not important; rather, it means that one of the three seems to receive more attention than the others. This might be related to the political context in the different cities: Montreuil has a vivid history of civic engagement and a strong communist legacy, whereas in the Japanese context seniority and hierarchy are important. The question is of course in how far these underlying values and worldviews are actually brought to the table and made a subject of a transition management process. This point nicely illustrates one of the transition governance principles put forth by Loorbach (2010, p. 167), that "*the dynamics of the system create feasible and non-feasible means of steering.*" With worldviews and values being part of system dynamics, they make a focus on one of the three dimensions (ideas, practise, or relations) more feasible.

A second general observation is that all five case studies focus on what was, is or will be, built in terms of ideas, practices, or social relations. None of the case studies details what needs to be 'broken down' in the current status quo. If one takes building up and breaking down as interrelated processes, the latter does deserve more attention (cf. Loorbach 2014).

A final general observation is that only in the case studies that persisted for longer periods of time can we observe the impact on taking up or mainstreaming ideas co-created in TM arenas, as only those case studies allow to position TM processes in relation to the larger planning processes landscape. While both Japanese case studies do cover longer timeframes and allow observing the mainstreaming of previously alternative ideas, practices and social relations, this is not the case for the European cases, which cover periods of less than 3 years. As such, an open question remains whether these ideas, practices, and social relations stabilise, formalise, or become mainstream.

9.4.1 Ideas

In terms of the generation of new ideas, I inductively distinguished and discuss the following: (1) visions and perspectives, (2) increased understanding of problems and solutions, and (3) insights and understanding about the nature of the process and (4) the nature of the knowledge and understanding created.

First, new knowledge is developed with regard to the future of all cities. The transition arena leads to the articulation of alternative visions and perspectives on the future of the city, as such helping participants to break loose of the 'dictatorship of the present.' The importance of such visions are articulated by Frantzeskaki and Tefrati (Chap. 4, this volume) in their chapter on Aberdeen: a "*vision acts as an attractor or mobilizing icon for networks to collaboratively break-through current patterns of practice that hold innovation latent.*" They manifest in vision documents that are disseminated across the cities.

Second, the development of alternative ideas also relates to new knowledge, expertise, and understanding with regard to persistent problems. Specific pathways and solutions were developed and written down in a transition agenda (for Aberdeen and Ghent), or as solution cards (in Montreuil). With the broadening of the problem frame, these ideas covered a large array of themes in Ghent, such as the importance of water cycles and the potential of city greening. In Kitakyushu, the result was the establishment of a recycling factory along the then-popular "*idea of completely eliminating waste and emissions by utilizing them as resources*" (Shiroyama and Kajiki 2016, Chap. 7, this volume).

Third, insights and understanding emerged with regard to how cities can work on sustainable development, and thus the design of engagement processes. Characteristics of successful processes are an open character, collaboration between different actors, learning from each other (e.g., municipality and citizens or cross-departmental collaboration), and a redefinition of hierarchy as in having a more 'horizontal' process. In Aberdeen, all participants of the transition management process saw a benefit in 'working together,' and in Ghent the lessons about 'co-creative approaches' stimulated the redefinition of roles (e.g., of the municipality, the administration) and relations (e.g., between citizens and municipality). This development is amplified by the fact that these processes involve at least three types of interactions: between the policy officers from different departments in the transition teams; between the policy officers and transition arena participants, and finally among transition arena participants. In Montreuil, for example, an emphasis was on interviews held by policy officers with inhabitants, which led to more knowledgeable civil servants. The transition management process was considered more horizontal and open than other participation frameworks, which has contributed to an experimentation with new working routines. The chapter on Ghent emphasises learning as an important aspect: "*knowledge of participants increased and complexity of the subject is acknowledged*" (Hölscher et al. 2016, Chap. 6, this volume).

Fourth, this new knowledge on future perspectives and hands-on solutions is of a different kind. On the one hand, the new knowledge emphasises a more systemic and holistic kind of thinking, and on the other a collaborative aspect thereof in terms of learning from other people's perspectives. With regard to systemic and holistic thinking, an important element of the alternative visions developed in the European cities is a broadened understanding of sustainability. These visions emphasised social, cultural, and political dimensions next to ecological aspects, making value discussions part of the process rather than ignoring them. Thus, the visions resulting from the arena form a distinct alternative to the institutional discourses with their strong focus on climate mitigation and energy savings.

9.4.2 Practices

In all cities, one can trace alternative practices triggered by transition management processes. For one, there are the co-creative practices of the transition arena. Second, there are the many experiments and projects that were started in the cities to address the unsustainability challenges.

All cities have been engaging in new practices of collaboration; this holds for the European cities with an operational application of transition management but also for the Japanese cities. In Kitakyushu, interesting practices of liaising and bridging between local and national government officials can be observed, as well as training of government officials to increase the interaction between them and the broader public. In Higashiomi, community business models were established as a practise where different actors and funding streams come together. In the European cities, municipalities were in the lead of implementing transition management processes that proved to be alternatives for current participatory practises, caused, amongst others, by the open agenda, the co-creative nature, and the use of new facilitation techniques (e.g., in Montreuil). It also initiated a change from working in silos to allow for cross-departmental exchange and pollination within city administrations (e.g., Montreuil, Ghent). This practice of collaboration extends to include citizens and policy makers and at times also business actors in different forums, such as the transition team, and transition arena, but also in working groups or projects.

Next to these more general observations, all cities have started to address sustainable development through a number of experiments and projects that are very specific to each context. By way of example, in Higashiomi, multiple niche practices in the areas of food, energy, and healthcare were combined in one physical location, the Welfare Mall. The city of Ghent saw a number of working groups each focusing on a different topic, some of which were more successful in also establishing projects or practices, such as a 1-day carrot mob or a very successful 'living streets' project in which citizens temporarily turn their street into a no-go area for cars. In Kitakyushu, the PET bottle recycling business was established as a new successful practise of local business rejuvenation.

9.4.3 Social Relations and New Actors

As part of a transition management process, new and/or alternative social relations and actors (roles) are developed and nurtured. All cities show a change in social relations, although in different forms: (1) creation of new relations, (2) change of the quality of relations, (3) change of existing roles, and (4) creation of new actors and/or roles. Roles are not static, but are in constant development, and different actors can assume different roles in different contexts. An increase in reflexivity helps to become aware of such ongoing negotiations of roles and relations. In the transition arena processes in the European cities, 'external' actors, such as transition researchers (in Ghent and Aberdeen) or the Local Energy Agency (in Montreuil), took on the task of increasing the reflexivity of the group and therewith the quality of the process and its outcomes.

The creation of new relations can most easily be outlined on the level of individuals: the transition arena participants established new ties with one another, the policy makers from different departments established cross-departmental ties with colleagues, and individuals from both groups established ties with one another. The nature of these ties is not examined in detail in the individual city chapters; by way of example, the Montreuil case describes collaboration; the Ghent chapter gives insights into the temporariness of some of the ties (e.g., between transition arena participants); and the Kitakyushu example shows the longevity and formality of some of these ties. In the Japanese cities, these new ties include ties between individuals from government and business. In Aberdeen, for example, it seemed difficult to establish ties with the oil and gas industry.

As has become clear from discussing alternative ideas and practices, these do have repercussions for how actors relate to one another. Rather than being about participation of citizens in municipality-led processes, these processes are about being involved as citizen and policy maker in societal processes, where both actor groups do have different perspectives, stakes, and powers but are motivated by a 'shared' goal: sustainable development. More specifically, in co-creative processes, the relations between 'the government' and 'the citizens' is of a different nature and quality than those in the usual participatory processes (especially in citizen consultation processes). In Montreuil, this relation has been formalised in a charter which outlined that the relation was to be of horizontal nature (rather than hierarchical), based on "*collegiality, mutual respect, consensus, zero criticism principle and the common ownership of ideas elaborated*" (Krauz 2016, Chap. 8, this volume). Such a change in relations also questions the understanding of the underlying role. A policy officer interacting with citizens and interested stakeholders in a specific way will add this activity to his/her role understanding of what it is a policy officer should or should not be doing. As put for the case of Aberdeen: "*The process itself enabled them to identify a change of roles of the participants and the council to work towards the desired future in terms of having the city facilitating further activities*" (Frantzeskaki and Tefrati 2016, Chap. 4, this volume). Such changes in specific role understandings have repercussions for the understanding of other, related roles. As the understanding of what a 'policy officer' is and is supposed to do changes, related roles such as 'citizen' or 'the municipality' change with it (Turner 1990).

The same mechanism holds for the entry of new 'actors' or 'roles' in an existing network. I argued earlier that the transition team and the transition arena can be regarded as both settings and actors. Regarded as the latter, they are 'new' actors in a system that is becoming part of the complex and dynamic web of relations; as such, new actors provide an impulse to redefine roles and relations. Another example is the formalisation of the transition arena into a Local Transition Council in Montreuil, which makes the transition arena not only a temporary impulse but rather an actor that is here to stay and play a role in the social and political fabric of the city. In the Kitakyushu case, this new actor is a new company, the West Japan PET Bottle Recycling Co. Ltd. NPR, which was established based on the investment from five private enterprises and the city.

9.5 Lessons for Transition Governance

After having focussed on the context as well as the outcomes of transition management processes, this section outlines lessons for the operational process design (thus mainly focussing on the lessons from the European cases) as well as the lessons for the principles of transition governance in cities.

9.5.1 Lessons for the Operational Process Design

Operational applications such as the European cases describe the prescriptive application of transition management for setting up a participatory sustainability transitions process. Based on the insights from the empirical chapters, the following lessons for designing such operational applications can be drawn.

Explore Productive Tensions in Hosting a Transition Management Process
City administrations do play an important role in the operational transition management processes: they initiated and (co-)funded, at times even facilitated them. Importantly, they are considered as being the driver of the process as well as the glue, holding the process together: they are looked at to ensure continuity and

follow-up. As outlined by Frantzeskaki and Tefrati (Chap. 4, this volume) for Aberdeen: a "*central role of the city council is to facilitate a follow up process for the change agents to proceed with the needed tasks for vision realization.*" Also, the Ghent case confirms this, in that a certain degree of steering from the city administration seemed to be necessary for sufficient motivation, collaboration, and follow-up. As such the city administrations do have to strike a balance between taking ownership of the process and leaving the process to others. They need to resist the wish for closure by themselves, their superiors as well as by the transition arena participants, while keeping the process running. They also need to ensure and foster reflexivity about roles, activities, and expectations towards one another, possibly involving a third intermediary party in the process. This important productive tension arises from the fact that such processes are first and foremost seen as societal learning processes rather than as usual public participation processes.

Think Carefully About Follow-Ups

Next to thinking about the role of the city administration in the follow-up, especially when it initiated the process, there are other ways to ensure that the visions are translated into action, projects, and experiments. Elaborating a strategy as well as creating an atmosphere of trust seems crucial in ensuring higher levels of follow-up activities (cf. Hölscher et al. 2016, Chap. 6, this volume). Essential is the involvement of the 'right' people (see also the next lesson). The experiences in Ghent and Montreuil showed that people who can link their professional life, background, or ambition to the experiment in question are more likely to stay involved. Others felt constrained by lack of financial but also time resources, as they were engaged next to their professional and private commitments. To find these 'right' people, Ghent explicitly organised a 'headhunting' event, and Montreuil a festival. Another crucial point is the translation of the transition agenda into viable business opportunities. In the Japanese cities, the development of a public–private partnership and a community business model were catalytic milestones. There are also other ways to think about follow-up: transition management is meant to be an iterative cycle, rather than a linear planning process. This concept includes that the process is meant to be repeated on a regular basis. In Montreuil, this has led to the formalisation of the transition arena in a Local Transition Council.

Involve Niche-Regime Players

Important for the transformative potential of the transition management process are ties with dominant institutions, although these are also the object of change and should not hinder radical visions and experiments. Again, a balance that is difficult to strike. The example of Ghent showed that the fostering of these ties were opportune at a stage when the overall vision was set and when the focus was on broadening networks and initiating actions. Through headhunting events, the transition arena and the transition team aimed to involve those change-minded actors of the regime with transformative power. The Ghent example also illustrates how to link the transition agenda with the policy agenda and the political agenda (i.e., upcoming elections were taken as a deadline for having the transition agenda published), which can lead to higher uptake of the ideas, while of course also bearing the danger of dilution. This is by no means an easy process: it asks time, energy, and plenty of communication. Both Japanese cases show the fruitfulness of involving businesses and market actors in the process, actors often considered to be more aligned with dominant unsustainable paradigms. Both cases also demonstrate that processes at city level can be accelerated by being tied to processes or actors at other governance levels, such as the national government.

Induce Reflexivity into the Process

As the operational transition management process is explicitly aimed at questioning the status quo and fostering alternatives, it does profit greatly from increasing its reflexivity. A transition management process does not take place in a vacuum, and this is where reflexivity can play an important role. This stage includes reflections on overall purpose and the role of the transition arena in the broader governance setting (e.g., legitimacy, relation to other governance actors) and the setting of the transition arena (e.g., status as frontrunner, kinds of outcomes to be achieved, continuity aimed for). Reflexivity does not need to obstruct action; rather, a transition management process knows periods of opening up issues and of closing them down to enable action (which is ideally followed by an 'opening up' again). The question of the status of the transition arena, which became so pressing in Montreuil, exemplifies this: the question was opened, including an implicit resistance to closing it down by a 'verdict' of the transition team alone. Rather, the issue was opened up and in a collaborative process closed down and agreed upon, perpetuated in a charter. The degree of reflexivity of the process profits on the one hand from the presence of 'outsiders,' such as the Local Energy Agency in Montreuil or transition researchers in Ghent and Aberdeen, and on the other from a feeling of ownership of the process by all involved.

Take One of Three Dimensions as an Entry Point

Of a more hypothetical nature is the observation that it might be beneficial to the process to focus on one of three types of outcomes as an entry point: ideas, practices, or social relations. This would mean a process focuses explicitly (rather than only implicitly) on envisioning and experimenting with one of the three, such as social relations in the broader context of a sustainability transition. As the three dimensions are intertwined, new social relations would always be accompanied with alternative ideas and practices. In choosing one of the three dimensions, the transition team (and possibly the arena) should take account of and play into the existing dynamics of the city: what is necessary for boosting sustainable development? This choice would lead to an in-depth examination of that specific issue, and it could help to manage expectations about the process and its transformative potential. The Aberdeen case showed that "*interviewed change agents did not experience missing actions as a demotivating factor*" (Frantzeskaki and Tefrati 2016, Chap. 4, this volume); as such, it was the elaboration of a new vision, the working together of actors from different backgrounds, and a reframing on how to see the future of the city that were the most valuable for the existing dynamics in Aberdeen. On the other hand, the process in Kitakyushu was not in the first instance based on a grand vision or idea; rather, it was an incremental process of combining new practices and people.

9.5.2 Lessons Learned for Principles of Transition Governance in Cities

Based on the analysis of the wealth of material in the five empirical chapters, one can conclude that the transition governance principles formulated by Loorbach (2010, 2007) hold and have not lost their importance. These principles include, amongst others, that content and process are inseparable, that governance processes should be oriented on the long term, that objectives of such processes should be flexible and adjustable at system level, and that creating spaces for actors to build alternatives (as analyzed in depth in previous sections) is crucial.

However, in the context of cities, I suggest that these principles can be complemented by the following ones.

1. For increasing their transformative potential, governance processes at city level need to relate to and/or engage with (different kinds of) developments, processes, and actors at other governance levels.

Such linkages to developments, processes, and actors at different governance levels can allow for spaces of interaction to be created, for example, through funding (e.g., through collaboration projects such as the EU-funded MUSIC project, through subsidies such as for the Welfare Mall in Highashiomi), through offering platforms (e.g., the networking of Japanese cities through national networks), or through knowledge exchange (e.g., as in the learning network in Higashiomi).

2. Navigating a complex system, such as a city, involves an iteration between periods of opening up and periods of closing down.

Transition management has been framed as a reflexive governance approach (Kemp and Loorbach 2006). Although reflexivity has not been specifically integrated in the transition governance principles, some of these principals are witness to this orientation (e.g., the adjustability and flexibility of objectives, or the focus on learning as pre-condition for change). By suggesting this principle building upon the notions of opening up/closing down as introduced by Stirling (2008), I aim to highlight this reflexive character more specifically and bring it to the center of attention. The value of such reflexivity was most obvious in opening up and closing down of the framing of the transition arena in Montreuil, but in other cities we also see such movements of opening up and closing down that could be more explicit. Such periods of opening up can also include questioning the status quo, challenging historically bounded perceptions, and brainstorming about a diversity of alternatives and futures.

3. Questioning and challenging actor roles and social relations allows for change in the social fabric of society as an important aspect of societal transitions.

By specifically focusing on actor roles and relations, this principle increases attention on the social innovations, thus the changes in social relations (cf. Franz et al. 2012; Moulaert et al. 2013) which are part and parcel of transition governance processes. If understood as proposing new ways of governing a complex system such as a city, transition management does imply new relations among the actors in the city. The cases in this book have all witnessed changes in the relation between different actors and in terms of how roles of actors are understood.

9.6 Concluding Summary

This chapter has offered a number of reflections on the experiences of five cities with transition management, focussing on one of the transition governance principles and its envisioned outcomes, namely, creating space for change to build up alternatives. The cities of Aberdeen, Ghent, Highashiomi, Kitakyushu, and Montreuil are examples for finding new models for local direct democratic practices, building upon concepts such as empowerment, self-organisation, and (temporary) arenas for dialogue and co-creation.

This chapter pointed to the importance of local context and its dynamics. It deepened the crucial idea of spaces for interaction and learning, which are deemed necessary for alternative ideas, practices, and social relations to arise. Finally, it drew lessons for the operational application of transition management, as well as suggested a number of additional governance principles for transition management in cities.

References

Frantzeskaki N, Tefrati N (2016) A transformative vision unlocks the innovative potential of Aberdeen city, UK. In: Loorbach D, Wittmayer J, Shiroyama H, Fujino J, Mizuguchi S (eds) Governance of urban sustainability transitions, pp 53–72. Springer, Tokyo

Frantzeskaki N, Loorbach D, Meadowcroft J (2012) Governing transitions to sustainability: transition management as a governance approach towards pursuing sustainability. Int J Sustain Dev 15(1/2):19–36

Franz HW, Hochgerner J, Howaldt J (2012) Challenge social innovation: potentials for business, social entrepreneurship, welfare and civil society. Springer, Heidelberg

Hölscher K, Roorda C, Nevens F (2016) Ghent: fostering a climate for transition. In: Loorbach D, Wittmayer J, Shiroyama H, Fujino J, Mizuguchi S (eds) Governance of urban sustainability transitions, pp 95–115. Springer, Tokyo

Kemp R, Loorbach D (2006) Transition management: a reflexive governance approach. In: Voß J-P, Bauknecht D, Kemp R (eds) Reflexive governance for sustainable development. Edward Elgar, Cheltenham/Northampton, pp 103–161

Krauz A (2016) Transition management in Montreuil: towards perspectives of hybridisation between 'top-down' and 'bottom-up' transitions. In: Loorbach D, Wittmayer J, Shiroyama H, Fujino J, Mizuguchi S (eds) Governance of urban sustainability transitions, pp 137–154. Springer, Tokyo

Loorbach D (2007) Transition management. New mode of governance for sustainable development. PhD thesis, Erasmus University Rotterdam, Rotterdam

Loorbach D (2010) Transition management for sustainable development: a prescriptive, complexity-based governance framework. Governance 23(1):161–183

Loorbach D (2014) To transition! Governance panarchy in the new transformation. Inaugural lecture. Erasmus University of Rotterdam, Rotterdam

Loorbach D, Shiroyama H (2016) The challenge of sustainable urban development and transforming cities. In: Loorbach D, Wittmayer J, Shiroyama H, Fujino J, Mizuguchi S (eds) Governance of urban sustainability transitions, pp 3–15. Springer, Tokyo

Loorbach D, Frantzeskaki N, Huffenreuter LR (2015) Transition management: taking stock from governance experimentation. J Corp Citiz 58:48–66

Mizuguchi S, Ohta K, Beer PJ, Yamaguchi M, Nishimura T (2016) Interactions among multiple - niche-innovations and multi-regimes: the case of the "Welfare Mall" in Higashiomi. In: Loorbach D, Wittmayer J, Shiroyama H, Fujino J, Mizuguchi S (eds) Governance of urban sustainability transitions, pp 73–93. Springer, Tokyo

Moulaert F, MacCallum D, Mehmood A, Hamdouch A (eds) (2013) The international handbook on social innovation: collective action, social learning and transdisciplinary research. Edward Elgar, Cheltenham/Northampton

Nevens F, Frantzeskaki N, Loorbach D, Gorissen L (2013) Urban Transition Labs: co-creating transformative action for sustainable cities. Journal of Cleaner Production. 50:111–122

Nevens F, Roorda C (2014) A climate of change: A transition approach for climate neutrality in the city of Ghent (Belgium). Sustainable Cities and Society 10:112–121

Roorda C, Wittmayer J, Henneman P, van Steenbergen F, Frantzeskaki N, Loorbach D (2014) Transition management in the urban context, Guidance manual. DRIFT/Erasmus University Rotterdam, Rotterdam

Shiroyama H, Kajiki S (2016) Case study of Eco-town Project in Kitakyushu: tension among incumbents and the transition from industrial city to green city. In: Loorbach D, Wittmayer J, Shiroyama H, Fujino J, Mizuguchi S (eds) Governance of urban sustainability transitions, pp 117–136. Springer, Tokyo

Stirling A (2008) "Opening up" and "closing down." Power, participation, and pluralism in the social appraisal of technology. Sci Technol Hum Values 33(2):262–294

Turner RH (1990) Role change. Annu Rev Sociol 16:87–110

Van Buuren A, Loorbach D (2009) Policy innovation in isolation. Conditions for policy-renewal by transition arenas and pilot projects. Public Manag Rev 11:375–392

Wittmayer J, van Steenbergen F, Quist J, Loorbach D, Hoogland C (2011) The community Arena: a co-creation tool for sustainable behaviour by local communities. Methodological guidelines. Deliverable 4.1, InContext: EU ENV.2010.4.2.3-1 grant agreement n° 265191. Online available here: http://www.incontext-fp7.eu/sites/default/files/D4-1_Methodological%20guidelines_final.pdf

Wittmayer JM, Schäpke N, van Steenbergen F, Omann I (2014) Making sense of sustainability transitions locally. How action research contributes to addressing societal challenges. Crit Policy Stud 8(4):465–485

Wittmayer JM, van Steenbergen F, Rok A, Roorda C (2015) Governing sustainability: a dialogue between Local Agenda 21 and transition management. Local Environ: Int J Just Sustain. doi:10.1080/13549839.2015.1050658

Wittmayer J, Mizuguchi S, Rach S, Fujino J (2016) City networks for sustainability transitions in Europe and Japan. In: Loorbach D, Wittmayer J, Shiroyama H, Fujino J, Mizuguchi S (eds) Governance of urban sustainability transitions, pp 37–50. Springer, Tokyo

Chapter 10
Practical Recommendations for Policy Makers and Practitioners for the Governance of Urban Sustainability Transitions

Pieter J. Beers

Abstract With increasing complexity, societal issues cannot be managed using the classical "rules and regulations" associated with traditional government. Rather, they require more subtle, network-oriented arrangements such as transition management, in which policy makers are not the only partners, and not necessarily the leading partner either. This chapter takes stock of the experiences with five urban sustainability challenges from Japan and Europe to draft recommendations for practitioners to more effectively handle such issues. It also reflects on examples of when and why to use transition management specifically based on these experiences. These examples show the importance and potential of network *hybridisation*, in the broadest sense; network hybridisation has sectoral, administrative, niche/regime, and grassroots/incumbent dimensions, all of which can provide opportunities for transition. Regarding transition management, we learnt that information provided by scientists can act as a *common starting point* in arenas with diverse participants, who can use it to connect their own context and practice to the arena issue at hand. Furthermore, these cases suggested that striving for *shared actions* and connecting different problem orientations is more fruitful for transitions than striving for *consensus*.

Keywords Sustainability transitions • Urban sustainability • Governance recommendations • Transition management • Reflexive governance • Networks • Network hybridisation

10.1 Introduction

A growing awareness exists that, with increasing complexity, societal issues cannot be managed using the classical "rules and regulations" approach associated with traditional government. Rather, they require more subtle, network-oriented

P.J. Beers (✉)
Dutch Research Institute for Transitions, Erasmus University Rotterdam, Rotterdam, The Netherlands
e-mail: beers@drift.eur.nl

D. Loorbach et al. (eds.), *Governance of Urban Sustainability Transitions*, Theory and Practice of Urban Sustainability Transitions,
DOI 10.1007/978-4-431-55426-4_10

arrangements in which policy makers are not the only partners, and not necessarily the leading partner. Policy makers need to collaborate with businesses, scientists, NGOs, citizens' groups and initiatives, and partners from education to stumble upon possible solutions, as can be learnt from the recent calls for reflexive governance by Voß and Kemp (2006) and Grin (2006).

Approaches for reflexive governance such as transition management (Loorbach and Rotmans 2006) are relatively new and have primarily been implemented and tested in specific and local contexts, despite their importance for the grand challenges of this age. Furthermore, the associated tools and methods are still in development (true to their reflexive nature) and subject to change. Hence, the importance to take stock of recent experiences with transition management and similar examples of policy processes in the context of complex societal challenges, and to derive lessons for practitioners so that they more effectively may work towards societal change, is evident.

The studies collected in this book reflect two types. The European studies of Aberdeen (Frantzeskaki and Tefrati 2016, Chap. 4, this volume), Ghent (Hölscher et al. 2016, Chap. 6, this volume), and Montreuil (Krauz 2016, Chap. 8, this volume) all represent applications of transition management, whereas the studies in Higashiomi (Mizuguchi et al. 2016, Chap. 5, this volume) and Kitakyushu (Shiroyama and Kajiki 2016, Chap. 7, in this volume) from Japan are more free-form examples of governance for complex societal problems, yet they are analysed using a transition management framework. The main difference between the European and Japanese chapters is that the European chapters focus on a transition management intervention, giving a detailed account of a relatively short period, whereas the Japanese chapters report on 'complete,' mostly historical, transitions.

In this chapter, we gather both explicit and entailed governance recommendations based on the empirical chapters and their associated governance challenges. These recommendations are specifically aimed at all professionals who practically contend with urban sustainability issues: these include urban planners and policy makers, who through our recommendations can benefit from lessons learnt elsewhere. For academics, especially action researchers, the recommendations in this chapter can be seen as heuristics to increase the effectiveness of a scientific contribution to processes of social change and as points of concern for gathering research data about urban sustainability issues and the associated governance processes.

We distinguish between two types of recommendations, general governance recommendations and recommendations for transition management approaches (Fig. 10.1). First, the general recommendations for the governance of complex societal problems, gathered from all the empirical chapters, should be of interest to anybody looking for insights about the often messy but also inspiring practice of working towards sustainability. Second, we move to recommendations specific for the use of transition management approaches. These recommendations are not intended as solutions, but rather as guidelines/heuristics that practitioners may use to further transition efforts.

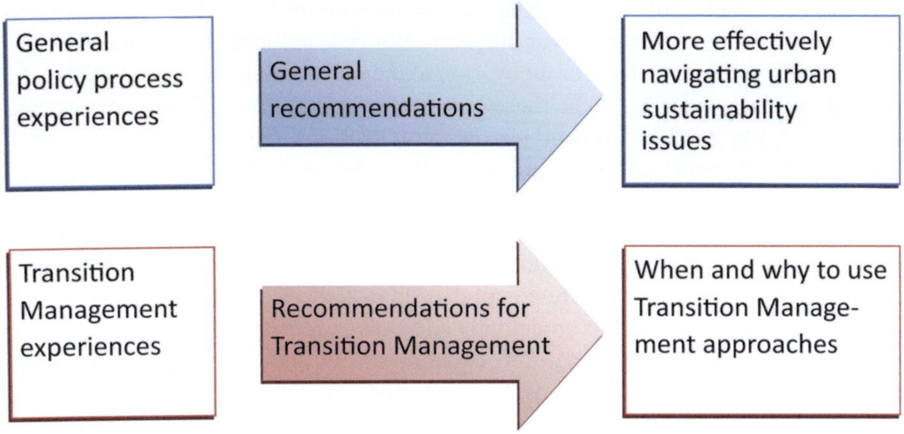

Fig. 10.1 Two types of recommendations

10.2 General Governance Recommendations

The general governance recommendations basically cover two broad types. The first set of recommendations very broadly concern (expectations about) time and action (Sects. 10.2.1 and 10.2.2). In a sense, they reflect the tension in transitions between concrete short-term actions and keeping with long-term visions of sustainability. The temporal aspect of transitions is key here. The second set of recommendations concerns the complexities of local and multilevel networks and context (Sects. 10.2.3 and 10.2.4). Both the multilayered nature of transitions and their geospatial characteristics here indicate the necessity of flexibility in the governance for sustainability, and they underscore that governance recommendations should not be taken to be more than heuristics.

10.2.1 Raise Realistic Expectations About Short-Term Actions

The one thing clear from all cases in this book is that transitions take time. This observation may come across as a platitude, especially given that definitions of transition often include a time horizon of one or even more generations. And yet, from a practitioner's point of view it is important to have (and raise) realistic expectations for concrete effects of transition efforts and impulses in the short term.

The Japanese cases all show long timelines. For instance, the Higashiomi case arguably has 40-year-old roots in antipollution activism and associated community-based business development using local resources. The novelties that together culminated in the Higashiomi Welfare Mall were 10 years in the making, and the

Welfare Mall itself took 5 years from the first meeting to be established. Similarly, the Kitakyushu case stretches over 30 years. Both cases illustrate that it can take a long time from the first ideas for innovation to the first actions, the first spades in the ground, so to speak. As such, these cases also underscore the importance and duration of the predevelopment phases of transitions (Rotmans and Loorbach 2009), and that transitions come in leaps and bounds once the ground is prepared.

In contrast, the chapters about the European cases span short time periods of only a couple (1.5 to 4) of years. They were explicitly positioned within an historical analysis and thus embedded in a longer timeframe, but within the project scope by definition relatively few actual actions could be accomplished. Here it becomes clear that inherent to proactively addressing transitions is the balancing act between understanding the long-term dynamics while focusing on short-term and relatively small-scale interventions. In Ghent, for instance, one of the few concrete actions was a "carrot mob" that attracted 938 participants and led the targeted supermarket to invest €10.000 in sustainability measures. Such actions may seem modest on the transition scale, but more noteworthy is the long-term potential for change evolving from the processes, such as the opening up within the city administration, two more arena processes, and the further network development. Similar results were achieved in Montreuil, where few actual actions were established during the transition initiative there. However, the transition management approach did yield intangible results such as learning and changes in work routines. New collaborations have been started, such as between the local energy agency and frontrunners and between the council house office and frontrunners, trying to implement transition experiments involving the limiting factor of lack of funding.

It is interesting to also point to the initial skepticism that the transition management approach received in Ghent, from both the city administration and the involved arena participants, because of the tension between their 'ambitious' expectations and the long-term character of transitions. Although transition management is aimed at accelerating change, the experiences reflected in the Japanese cases suggest that, even with transition management, we should still be modest about how much action we can expect in a short time, and that efforts towards transitions require long-term commitments that may span several projects and arenas (perhaps temporary transition impulses) over time.

We can draw two immediate governance recommendations:

- Have realistic short-term expectations about transition impulses; be careful with creating expectations of action in the short term, and focus on the less tangible changes in the form of network growth and discourse change.
- Think beyond the project horizon; 'a few years' is only a moment on the time scale of a transition. Make a long-term commitment; otherwise, efforts might be wasted.

10.2.2 Work Hard Towards Short-Term Actions

The previous point underscores the importance of efforts towards actions. The European cases are indicative of the importance of finding resources that create connections to people with time, ideas, knowledge, skills, money, etcetera, to engage in concrete activities. In Ghent, "transition networks mainly involve actors who can link to a professional level and have the capacity of translating new ideas to new practices or 'selling' ideas to a broader audience" (Hölscher et al. 2016, Chap. 6, this volume). This point shows the importance of linking transition arena ideas to the professional contexts of transition arena members to result in action.

Similarly, the Montreuil case showed how hard it is to move towards actions, in this case because of limited availability of resources. The predominantly grass-roots-oriented transition arena aimed to include both Higher and Lower Montreuil, two areas with important differences in affluence and culture, but the resulting network consisted mainly of "white males," not offering the more equal representation that the initiative intended. The case shows that even only participating in an arena still requires time and resources that many people may not have at their disposal.

The examples of Ghent and Montreuil underscore both how difficult it is to come to action in the short term and the importance of finding resources for action. Of course, resources are not the only issue here. In Aberdeen and Montreuil commitment and support from the city administration was shown as equally important. In this regard, also more intangible aspects are relevant (mainly networks and motivation). Still, all examples show how hard it is to move towards action without resources.

In Kitakyushu, subsidies were used as an instrument to accelerate change, as also happened in Higashiomi, where subsidies were applied "flexibly," in the sense that sometimes the rules and resources for subsidies were changed in accordance with changing situations. This example shows that also the more traditional policy instruments of rules/regulation and financial instruments such as subsidies and fines can be used to create conducive conditions for new markets, which leads us to the following governance recommendation:

- Use policy to create conditions for new markets; create spaces for interests, values, and dreams to meet, and shift resources (time, money, power, ...) towards the niche.

The issue of how resources are necessary for joint action also points towards the role of increasingly hybrid networks. Transition initiatives need to create links to resources for action, suggesting two additional governance recommendations:

- Be aware of and sensitive to (the necessity of) (network) resources for action, and try to create new network links that can cater for action.
- If possible, create space for the unrepresented by sharing resources; this will benefit the local network of the initiative.

10.2.3 On the Topic of Network Hybridisation

Across the various cases in this book, the importance of diversity in networks (i.e., network hybridisation) comes to the fore in many different guises: spanning different geospatial levels, different levels of administrative scale, bridging between business, education, science, government, and NGOs, etcetera. We can define network hybridisation as a multidimensional form of diversity in networks, and each of the constituting dimensions (sectoral, administrative, niche/regime, grassroots/incumbents) represents opportunities for using differences for the benefit of transition.

The most typical form of network hybridisation would seem the combination of bottom-up and top-down steering: grassroots meeting government, niche versus regime. However, the cases in fact exhibit network hybridisation on many different dimensions, all seemingly fit to the specific case characteristics. Also, such hybridisation appears to be a more general and nondichotomous structure than the simple niche–regime distinction.

In the cases of Kitakyushu and Higashiomi, hybridisation occurred in terms of business–government collaboration that spanned both local and higher administrative levels. In Higashiomi, hybridisation appeared specific for Higashiomi's geographic location. Higashiomi is located between Nagoya, which has been an economic-industrial centre since medieval times, and Kyoto, the former (until 1868) capital of Japan and a current-day academic centre with a university that is internationally well regarded. Together, these neighbouring cities act as resources for hybridisation, for example, in terms of in-sourcing knowledge for cross-sectoral collaboration.

In Kitakyushu's Eco-town Project, close relations between city and national government officials were vital in negotiating for a very beneficial subsidy scheme that enabled local companies to cover both "soft" subsidies for project preparation and "hard" subsidies for infrastructure improvement. In that sense, these cases show that hybridisation is a multidimensional concept that can concern multiple administrative scale levels and also multiple sectors.

The Kitakyushu case is also illustrative for the niche–regime type of hybridisation, especially because it runs counter to the emphasis that has been put on the importance of supporting niche developments through grassroots initiatives. The actors involved in the Eco-town Project, such as high-level government officials and business executives from Nippon Steel and Mitsui & Co, can all be regarded as *incumbents*, powerful actors with important functions in the existing regimes. Although the Eco-town Project can arguably be seen as a niche, it certainly was not initiated by grassroots participants. Rather, it is a case of niche formation by incumbents that clash with the regime. The case shows that transitions can begin in many different places and ways.

In Ghent, we witnessed network hybridisation on a more abstract level, in the guise of a sense of mutual similarity and like-mindedness, regardless of the original motivations and concerns of the actors involved: this increased intrinsic motivation and a feeling of leverage. Another example from Ghent concerns the shift from collaboration between administrative university staff, students, and professors to a more comprehensive collaboration in transition arenas, that is, the transition UGent process. The same goes for hybridisation within the administration: cross-departmental linkages were created, most evident in the mobility arena. With regard to Montreuil and Aberdeen, the collaboration between the administration and transition arena participants constitutes another example of hybridisation, moving on from the importance of 'bottom-up' versus 'top-down' to more diverse forms of hybridization.

We can define two general lessons about network hybridisation.

- Try to keep an open eye out for opportunities on *all* dimensions (sectoral, administrative, niche/regime, grassroots/incumbents) instead of searching blindly for a grassroots initiative.
- For policy makers specifically, the cases show the importance of creating links between local and higher administrative levels to create further opportunities for strengthening transitions. If possible, try to establish such links yourself.

10.2.4 Be Sensitive to Local Conditions

A final lesson to be drawn from the examples mentioned here and elsewhere in the book is the necessity of being sensitive to local conditions. For instance, in the case of Ghent, there was a high policy focus on sustainability, and policy makers had the ambition to turn Ghent climate neural. They established a Climate Alliance to involve the actors necessary for achieving this ambition, which offers a very different starting point than the case of Montreuil. There, the grassroots orientation of the arena was in part the result of the legacy of 70 years of municipal communism, involving a strong structuration of civil society in associations and a strong participation of citizens in formal and informal (mainly former communist) networks (despite the election in 2008 of a mayor from a green party).

The Higashiomi Welfare Mall is also an example in case. The specific history and location of Higashiomi may actually have benefited making connections between different frames of reference (Nagoya versus Kyoto) and creating value for multiple perspectives (Regeer et al. 2011; "dosho-imu," or "connected value development"). Specifically, a local sense of identity as descendants of the 'Ohmi-Merchant' may have fostered a sense and skill for connecting different values. The 'Ohmi-Merchant' is regarded as a prototype of the modern Japanese merchant, known for their pragmatic philosophy, that business is good when seller, buyer, and society all see it as good, a philosophy that can be understood as a direct plea for connecting different values. Thus, local conditions offer specific barriers and, more importantly, specific opportunities for policy makers to foster and accelerate efforts towards transition.

The above examples about local conditions lead to additional governance recommendations about network hybridisation.

- Do not limit network hybridisation to combining "top-down" with "bottom up" developments.
- Keep an open eye for any possible way of hybridisation that may offer new resources for change.

10.3 Recommendations for Transition Management Approaches

In this section, we gather recommendations, derived from specific aspects of transition management, that may inform future transition management initiatives and which might also be followed in sustainability governance processes in general.

We base ourselves on the empirical examples in this book of how transition management (and/or its associated tools and instruments) influences and accelerates transition efforts. The cases uncovered some considerations for using transition management in specific contexts. In this section, we focus on common starting points for transition efforts and on the role of transition management for the acceptance of diversity.

10.3.1 Transition Management Can Offer a Common Starting Point

The transition management processes in all three European cities started with a system analysis. The lessons from the Ghent case suggest that this helped to ensure a level of realism and attainability to the transition arena. In Montreuil, the system analysis was also largely accepted as a good starting point for a discussion. It relied on insights proposed by frontrunners during their interviews, complemented by desk research by the transition team: this ensured its acceptability. The ensuing discussion focussed on the framing of the transition challenges. These result are surprising (and encouraging), especially in light of increased politicisation of policy processes, as well as past criticisms holding that transition management is too directive and top-down to function in the societal complexities of the real world (Shove and Walker 2007).

Specifically, both the system analysis and the future envisioning process helped the participants in all three cities to create shared understandings and to facilitate dialogue and co-creation. In the system analysis, the researchers worked together with Ghent municipal policy makers to assess stocks and interrelations of social, ecological, and economic value (Grosskurth 2008). In the subsequent first arena meeting, the results of the system analysis were used as input for a broader discussion that helped the arena participants to draw connections between their own perspectives and the topic of climate change.

In general, the information compiled in a system analysis is not just put together from scientific sources and distributed in written form. Rather, it was fed into the process and discussed with the participants in all three cities. The resulting problem orientation was accepted by the participants as theirs. In other words, the arena discusses the information to develop its own (possibly multiple) perspective(s). The information becomes seen as largely neutral through a social process of learning, not because 'researchers state that this is how things are,' so to speak.

Furthermore, both the Ghent and Montreuil cases suggest that such information should not be sourced from science alone. Non-scientists' involvement in drawing up the requirements of the system analysis may have been vital to the approach, as well as not offering it as 'truth' but only as inspiration for the envisioning process. Distributing and discussing information from scientists is inherent to a transition management approach, but the experiences gathered here suggest that doing so can also be useful for governance issues in general when participants with multiple perspectives need to collaborate. We can support the following governance recommendations:

- Use knowledge and information from scientists to contribute to a *common starting point* for participants in diverse policy/transition arenas. They can use such information to connect their own context and practice to the arena issue at hand.
- Create space for reframing and reinterpreting knowledge and dominant (policy) frames by engaging with different societal perspectives and types of knowledge.

10.3.2 Transition Management Can Help to Make Diversity More Accepted

In Aberdeen, transition managementprovided a different function than the foregoing: it created space for the coexistence of mutual differences in the arena, and as such provided a basis for creating connections between different perspectives (Beers et al. 2010). Specifically, the envisioning process in Aberdeen was of tactical value for uncovering specific current practices that stand in the way of transformative change, which also enabled identifying options for collaborative breakthrough actions. In this capacity, the envisioning process revealed the potential of creating ties with stakeholders that the initial arena participants usually would not consider, in other words, the necessity of gathering unlikely allies, up to and including the oil and gas industries. In Ghent, something similar happened when new actors were invited to join individual climate working groups as a follow up of the climate arena.

Being aware of their mutual differences, the actors in Aberdeen searched for synergies, for instance, change agents sought after possibilities for concerted action with city projects, despite the hindrances that the current municipal administrative system presented. The envisioning processes helped participants become aware of how they were connected to others, despite having different concerns and aims. This process appears akin to "connected value development" / "dosho-imu," in the sense that shared concerns are not a necessity for shared actions (of course, they do help). Rather, actions must be able to *connect* different concerns, by having valuable and specific results for each participant. Apparently, the Aberdeen envisioning process, by way of political empowerment, helped search for connections for shared actions amid different perspectives.

It must be noted about the envisioning process in Aberdeen that the stakeholders involved concerned mainly government officials and change agents. These two groups certainly shared (somewhat) opposing views, in the sense that administrative institutions often presented obstacles to the goals of the change agents. However, the actual difference in interests and values might have been bigger still if, for instance, the oil industry had also been part of the envisioning process. In that light, the above result should still be treated as rather context specific, although its indications for the potential of envisioning processes as part of transition management are certainly promising.

Multiple, different perspectives are part and parcel of any sustainability governance processes, and not specific for transition management. The cases discussed here suggest that doing justice to these differences can still serve as a basis for joint action. Rather than striving for consensus, that is, an agreement about problem orientations and future visions, it appears important to search for shared actions that are meaningful and beneficial even from different perspectives. Two governance recommendations are implied:

- Search for shared actions more than for consensus on goals and specific targets.
- Use future envisioning processes to help bridge different perspectives in the search for shared actions.

The above recommendations allude to two more general characteristics of transition management in terms of empowerment (cf. Hölscher et al. 2016, Chap. 6, this volume). First, it can act as a type of *cognitive empowerment* that helps to distinguish (in practice, at least) between what is accepted as "neutral" and what is treated as opinion. Second, it can act as a bridging device that can connect multiple agendas/frames of reference and in so doing support *political empowerment* of the transition arena.

10.4 Conclusion

This chapter derived various governance recommendations from the European and Japanese cases of transition and transition management collected in this book. We discussed recommendations based on the tension between short-term and long-term aspects of transitions and others concerned with the multilayered and geospatial aspects of transitions. With regard to the former, we recommend being careful about which expectations to raise about the impacts of a transition impulse. In the short term, it might be better to focus on the intangible results of network growth and discourse change than on concrete tangible actions. One transition impulse is only a short moment on the time scale of an overall transition. An associated suggestion is to think beyond the time horizons of individual projects and to make a long-term commitment.

Although concrete action (implementation of innovations, etc.) might be difficult to achieve in the short term, actions are also an important aspect of learning towards transitions. Therefore, a second set of recommendations concerns how to accelerate towards local actions on the short term. Networks are key here, as connections of different types of resources (time, money, power, etc.). Policy makers can contribute to action-oriented change by creating space for different actors to meet and shifting resources to niches. Those without resources, who may remain unrepresented in a transition initiative, may require specific attention in this regard.

This point leads us to the multilayered and geospatial specifics of transitions. The cases suggest keeping an eye open for network *hybridisation* opportunities, which should not be limited to combining "top-down" with "bottom-up" developments. Rather, networks have sectoral, administrative, niche/regime, and grassroots/incumbents dimensions, which all can provide opportunities for network hybridisation. Traditionally, transitions have often been thought of as originating from grassroots initiatives, whereas these cases have shown that incumbents may also work towards transition (Geels and Schot 2007). This concept of network hybridisation in fact suggests that initiatives for transition might begin in many different ways, from many different actors. The challenge then is to keep an open eye for any possible way of hybridisation that may offer new resources for change.

For when and why to use transition management, we have learnt that information provided by scientists can act as a *common starting point* in arenas with diverse participants, who can use it to connect their own context and practice to the arena issue at hand. Furthermore, these cases suggested that striving for *shared actions* that connect different problem orientations is more fruitful for transitions than striving for *consensus*.

In closing, perhaps the most notable conclusion is the importance of local context, in terms of obstacles and opportunities for change. This realization underscores the necessity of using transition management as a set of guidelines, not as a 'cook book.' The recommendations collected here suggest that transition management can be seen as a balancing act between these points:

- Top-down versus bottom-up
- Incumbents versus grassroots
- Expectations versus actions
- Inclusion of underrepresented perspectives versus inclusion of time, money, and resources for action

In that light, it is very difficult to draw concrete governance recommendations beyond what we presented here. Still, the results from the European transition management cases indicate the potential of transition management for both cognitive and political empowerment, even in full awareness of the balancing acts entailed by transition management. In that sense, this book in fact suggests that transition management may actually be more complex but also more promising than it seems.

Acknowledgements Thanks to the reviewers for their useful comments, which greatly helped in improving this chapter.

References

Beers PJ, Veldkamp A, Hermans F, Van Apeldoorn D, Vervoort J, Kok K (2010) Future sustainability and images. Futures 42:723–732

Frantzeskaki N, Tefrati N (2016) A transformative vision unlocks the innovative potential of Aberdeen City, UK. In: Loorbach D, Wittmayer J, Shiroyama H, Fujino J, Mizuguchi S (eds) Governance of urban sustainability transitions, pp 53–72. Springer, Tokyo

Geels FW, Schot J (2007) Typology of sociotechnical transition pathways. Res Policy 36:399–417

Grin J (2006) Reflexive modernisation as a governance issue, or: designing and shaping *re*-structuration. In: Voß J-P, Bauknecht D, Kemp R (eds) Reflexive governance for sustainable development. Edward Elgar, Cheltenham

Grosskurth J (2008) Regional sustainability. Tools for integrated governance. Maastricht University, Maastricht

Hölscher K, Roorda C, Nevens F (2016) Ghent: fostering a climate for transition. In: Loorbach D, Wittmayer J, Shiroyama H, Fujino J, Mizuguchi S (eds) Governance of urban sustainability transitions, pp 95–115. Springer, Tokyo

Krauz A (2016) Transition management in Montreuil: towards perspectives of hybridisation between 'top-down' and 'bottom-up' transitions. In: Loorbach D, Wittmayer J, Shiroyama H, Fujino J, Mizuguchi S (eds) Governance of urban sustainability transitions, pp 137–154. Springer, Tokyo

Loorbach D, Rotmans J (2006) Managing transitions for sustainable development. In: Olsthoorn X, Wieczorek AJ (eds) Understanding industrial transformation: views from different disciplines, vol 44, Environment and policy. Springer, Dordrecht, pp 187–206

Mizuguchi S, Ohta K, Beers PJ, Yamaguchi M, Nishimura T (2016) Interactions among multiple - niche-innovations and multi-regimes: the case of the "Welfare Mall" in Higashiohmi. In: Loorbach D, Wittmayer J, Shiroyama H, Fujino J, Mizuguchi S (eds) Governance of urban sustainability transitions, pp 73–93. Springer, Tokyo

Regeer B, Mager S, Van Oorsouw Y (2011) Licence to grow. Innovating sustainable development by connecting values. VU University Press, Amsterdam

Rotmans J, Loorbach D (2009) Complexity and transition management. J Ind Ecol 13(2):184–196

Shiroyama H, Kajiki S (2016) Case study of Eco-town Project in Kitakyushu: tension among incumbents and the transition from industrial city to green city. In: Loorbach D, Wittmayer J, Shiroyama H, Fujino J, Mizuguchi S (eds) Governance of urban sustainability transitions, pp 117–136. Springer, Tokyo

Shove E, Walker G (2007) Caution! Transitions ahead: politics, practice, and sustainable transition management. Environ Plan A 39:763–770

Voß J-P, Kemp R (2006) Sustainability and reflexive governance: introduction. In: Voß J-P, Bauknecht D, Kemp R (eds) Reflexive governance for sustainable development. Edward Elgar, Cheltenham

Chapter 11
Sketching Future Research Directions for Transition Management Applications in Cities

Niki Frantzeskaki and Hideaki Shiroyama

Abstract This chapter proposes new research directions for transition management, building on the discussions and insights from the different chapters in this book. The research directions are clustered along the different application types of transition management: theoretical, heuristic, and operational applications. In doing so, we point to new research ground for advancing the research on transition management specifically and the governance of urban sustainability (transitions) in general.

Keywords Transition management • Sustainability • Research agenda • Transformation • Frontrunners

11.1 Introduction

Implementing sustainability agendas is a complex task for cities. Even when policy officers and planners are well aware of sustainability values and are environmentally literate, implementing agendas can become problematic because of the focus on short-term barriers rather than actions and the disconnect between the long-term objectives and action. Transition management is a governance approach that creates connections between long-term goals and short-term action, in the form of the transition agenda as practicality, through experimentation. In this volume, cases of practicing transition management showed the potential it has in bringing sustainability vision into practice. The cases from Europe and Japan have been elaborated

N. Frantzeskaki (✉)
Dutch Research Institute for Transitions, Erasmus University Rotterdam, Rotterdam, The Netherlands
e-mail: frantzeskaki@drift.eur.nl

H. Shiroyama
The University of Tokyo, Tokyo, Japan
e-mail: siroyama@j.u-tokyo.ac.jp

D. Loorbach et al. (eds.), *Governance of Urban Sustainability Transitions*, Theory and Practice of Urban Sustainability Transitions,
DOI 10.1007/978-4-431-55426-4_11

focusing on two of the different applications of transition management, namely, operational and heuristic applications (Frantzeskaki et al. 2014).

Before we proceed, we need to clarify that there is a distinction between governance of and for sustainability transitions. Governance of sustainability transitions refers to issues that are relevant to debate about types of interventions and how they play out in changing the direction and the pace of sustainability transitions. Governance for sustainability transitions is the prescriptive dimension: the methods, instruments, and frameworks that are promoting, enabling, or triggering sustainability transitions. Following this distinction, the theoretical and heuristic applications of transition management take up an interpretative way to analyse and examine past or contemporary transition dynamics. As such, they contribute to the understanding of the governance of sustainability transitions. The operational application of transition management provides knowledge for explaining the 'successes' or fitness of transition management as a process method that is instrumental to the governance for sustainability transitions. Transition management contributes to the governance for sustainability transitions. As sustainability transition scholars we invite more frameworks and methods to contribute to methods and frameworks for the governance for sustainability transitions.

11.2 Suggested Future Research on Governance of and for Sustainability Transitions

In this synthesis chapter, we distil future research directions for the development of transition management as a process methodology that creates transition governance 'situations,' the transition arenas, and more general future directions for research on governance for sustainability transitions. We use the three applications of transition management as indicated by Frantzeskaki et al. (2014) to structure our discussion.

11.2.1 Research Directions Inspired by the Theoretical Applications of Transition Management

Theoretical applications of transition management are included in Chaps. 1 (Loorbach and Shiroyama 2016, this volume) and 2 (Wittmayer and Loorbach 2016, this volume). In these chapters transition management is positioned as a new governance approach for enabling transformative change. Following the discussion of these chapters, we distil the following future research directions for the theoretical development of transition management:

- How can transition management methodology include methods to unpack and consider cultural diversity as a contextual element when applying it?

The consideration of culture and cultural diversity is a future research theme for the theoretical development of governance theory for sustainability transitions. Following the reflections of Meuleman (2013, pp. 46–47) on culture in governance for sustainability, and considering the chapters of this volume, we propose the following research questions to kick-start the scholarly debate on this theme:

- How does theory on governance for transitions consider cultural diversity in terms of diversity of values, beliefs, assumptions, and practises?
- How can theory on culture and cultural diversity inform governance for transitions?
- How can understanding and navigating cultural diversity enable us to understand transformation processes?
- What are new theoretical or conceptual frameworks that can inform a culturally responsive governance for sustainability transitions beyond considering culture a hindrance?

11.3 Research Directions Inspired by the Operational Applications of Transition Management

Operational applications of transition management are included in Chaps. 4 (Frantzeskaki and Tefrati 2016, this volume), 6 (Hölscher et al. 2016, this volume), and 8 (Krauz 2016, this volume). In these chapters transition management is applied as a process methodology for setting up participatory processes for scenario and strategy development. These processes can capitalize on the transformative and innovative capacities of actors, networks, and communities. Along with the discussions in these chapters, we distil the following future research directions for the operational development of transition management. We organized these along the different types of governance interventions suggested by Roorda et al. (2014).

Orienting

- How can transition management process design consider the insider–outsider actor dynamic of the transition management process so as to ensure accountability and legitimacy of the developed transition agenda?
- In which actor settings are transition arenas not applicable? What are the fitness criteria for the applicability of transition management as a governance process design?
- Which are the criteria for examining the fitness of transition management as a governance process design?
- Can transition management be applied in actor settings to infuse narratives on transformation in the absence of changemakers and frontrunning personal attributes?
- Can transition management be applied to enable, or empower transformative thinkers in policy, society, and business?
- What are the (pre)conditions for successful application of transition management?

Activating

- How can transition management processes and agendas contribute to social, economic, and environmental value creation?
- How can we assess the longitudinal impact of transition management agendas beyond the establishment of networks?
- What is the impact of transition management in creating social capital for transformations to sustainability?
- How to sustain and mobilise networks to take up and realize the transition agenda?

Agenda Setting

- How can the transition agenda be scaled down to local communities and local narratives without compromising or over-nuancing its transformative elements?
- How can processes of empowerment within the transition arena and across mobilized networks of actors be enabled and triggered?
- How to instrumentalise context conditions in mobilising action that aligns the transition agenda to ongoing governance programs and processes?
- How to align action suggested by the transition agenda with emerging/spontaneous action from bottom-up initiatives and networks?
- How to create conditions for coordinated synergy between top-down and bottom-up actions with the recommended transition agenda by the transition arena group?
- How can new value propositions from different theories and approaches (e.g., resilience approach, permaculture approach) be combined with a transition management approach at the agenda-setting phase?
- What are the actors to engage beyond frontrunners so as to realize the transition agenda? How to manage actors that require to shift roles and/or change skill-sets so as to partake in the transition?
- How to detect the new actor and actor–network setting that the transition agenda is proposing to prepare the actors for the changing roles and power relations?

Reflecting

- How are the specific methods that relate to each phase of the transition management cycle selected to fit the context complexity? (e.g., scenario, backcasting, system analysis)
- What are the different context-dependent definitions of frontrunners from the applications of transition management? How can we learn from these context-dependent definitions into systematising the frontrunners' selection methodology for transition management?
- How does issue framing relate to the identified and selected frontrunners for the transition management arena?
- How to evaluate the (degree of) radicality in a transition agenda in respect to current discursive elements in the case study context?
- How to evaluate the degree in which the transition agenda is culturally informed (inclusive versus exclusive narratives)?

- What are the means to share knowledge across transition management arenas? How to cross-compare cases considering the context conditions?
- How to understand the role and the different ways in which cultural diversity is represented in the transition agenda and transition vision?

11.4 Research Directions Inspired by the Heuristic Applications of Transition Management

Heuristic applications of transition management are included in Chaps. 5 (Mizuguchi et al. 2016, this volume) and 7 (Shiroyama and Kajiki 2016, this volume). In these chapters, transition management is employed as a descriptive or diagnostic lens to understand and explain dynamics of ongoing governance processes. Following the discussion of these chapters, we distil the following future research directions for the heuristic development of transition management:

- What are the controversies and conflicts that occur during the implementation of transition agendas between different actors as well as between the transition agenda and other issue or sector agendas?
- What are the similarities and differences of the transition management process designs across different contexts?
- What are the mechanisms of localising or scaling down the vision and ensuring its successful take-up by local networks and actors?
- What are the conditions of the institutional context that can facilitate the take up and anchoring of the transition agenda?
- What is the role of legal instruments in realising the transition agenda?

11.5 Challenges for Research on Governance of and for Sustainability Transitions and the Way Forward

Researchers who employ transition management have focused empirically on Western affluent societies that are locked in unsustainability. With very few cases of applications of transition management in non-Western contexts (e.g., case study work in South Pacific, India, Honduras, and Ghana in Frantzeskaki et al., forthcoming), the fitness of transition management in non-affluent societies and their governance culture can be a new ground for transition governance studies. It may therefore be relevant to combine different theoretical streams such as development studies with transition management as well as resource economics with transition studies to design new operational frames for transition governance overall.

Considering new cases and new contexts for applying transition management, we also wonder what it will take for transition management to be applied in cities facing sustainability threats such as 'shrinking cities,' 'bankrupt cities,' or neighborhoods in cities experiencing fast-paced transformation. In different contexts, transition management approaches may need to redesign and reconsider the actor selection process and or its criteria, or, more specifically, to move beyond frontrunners. What will be the characteristics of transformation builders or enablers in a context lacking frontrunners?

On a different note, the incorporation of findings from heuristic applications of transition management and from the interpretative research on transition governance into new frameworks and approaches for the governance for sustainability transitions is another direction. Unpacking and reflecting on past and contemporary transition processes and how they are governed can inform designs for governance interventions. So far these two streams of research have been neither coevolving nor contrasting. Therefore, we suggest incorporating findings from practising and reflecting to designing transition governance 'situations.'

In the same vein, researchers who apply transition management as an operational framework to design and set up participatory processes for strategic agendas have so far not enriched, adapted, or revised transition management tenets (Loorbach 2010) to further consider context dynamics. This feedback loop between operational applications of transition management and its theoretical grounds is still lacking: Chap. 9 (Wittmayer 2016, this volume), collecting insights and lessons on the governance of urban sustainability transitions, is a first notable exception. It may therefore be relevant to create a systematic monitoring and assessment framework that can assess the lessons learnt from the operational applications of transition management and contrast them with the theoretical grounds (tenets and process design of the transition arena). Such monitoring work will display operational design practices for transdisciplinary work aimed at enabling transformations.

References

Frantzeskaki N, Tefrati N (2016) A transformative vision unlocks the innovative potential of Aberdeen City, UK. In: Loorbach D, Wittmayer J, Shiroyama H, Fujino J, Mizuguchi S (eds) Governance of urban sustainability transitions, pp 53–72. Springer, Tokyo

Frantzeskaki N, Bach M, Hölscher K, Wittmayer J, Loorbach D (2014) Applications of transition management – taking stock of 13 years development pathway and identifying future research opportunities. International conference of sustainability transitions, Utrecht. www.ist2014.org

Frantzeskaki N, Hölscher K, Bach M, Avelino F (forthcoming) Co-creating sustainable urban futures, a primer on the application of transition management in cities. Springer

Hölscher K, Roorda C, Nevens F (2016) Ghent: fostering a climate for transition. In: Loorbach D, Wittmayer J, Shiroyama H, Fujino J, Mizuguchi S (eds) Governance of urban sustainability transitions, pp 95–115. Springer, Tokyo

Krauz A (2016) Transition management in Montreuil: towards perspectives of hybridisation between 'top-down' and 'bottom-up' transitions. In: Loorbach D, Wittmayer J, Shiroyama H, Fujino J, Mizuguchi S (eds) Governance of urban sustainability transitions, pp 137–154. Springer, Tokyo

Loorbach D (2010) Transition management for sustainable development: a prescriptive, complexity-based governance framework. Governance 23(1):161–183

Loorbach D, Shiroyama H (2016) Challenge of sustainable urban development. In: Loorbach D, Wittmayer J, Shiroyama H, Fujino J, Mizuguchi S (eds) Governance of urban sustainability transitions, pp 3–15. Springer, Tokyo

Meuleman L (2013) Cultural diversity and sustainability metagovernance. In: Meuleman L (ed) Transgovernance, advancing sustainability governance. Springer, Heidelberg/New York, pp 37–81

Mizuguchi S, Ohta K, Beers PJ, Yamaguchi M, Nishimura T (2016) Interactions among multiple niche-innovations and multi-regimes: the case of the "Welfare Mall" in Higashiohmi. In: Loorbach D, Wittmayer J, Shiroyama H, Fujino J, Mizuguchi S (eds) Governance of urban sustainability transitions, pp 73–93. Springer, Tokyo

Roorda C, Wittmayer J, Henneman P, van Steenbergen F, Frantzeskaki N, Loorbach D (2014) Transition management in the urban context. In: Guidance manual. DRIFT, Erasmus University Rotterdam, Rotterdam

Shiroyama H, Kajiki S (2016) Case study of Eco-town Project in Kitakyushu: tension among incumbents and the transition from industrial city to green city. In: Loorbach D, Wittmayer J, Shiroyama H, Fujino J, Mizuguchi S (eds) Governance of urban sustainability transitions, pp 117–136. Springer, Tokyo

Wittmayer J (2016) Insights and lessons for the governance of urban sustainability transitions. In: Loorbach D, Wittmayer J, Shiroyama H, Fujino J, Mizuguchi S (eds) Governance of urban sustainability transitions, pp 157–173. Springer, Tokyo

Wittmayer JM, Loorbach D (2016) Governing transitions in cities: fostering alternative ideas, practices and actors. In: Loorbach D, Wittmayer J, Shiroyama H, Fujino J, Mizuguchi S (eds) Governance of urban sustainability transitions, pp 17–36. Springer, Tokyo

Index

D. Loorbach et al. (eds.), *Governance of Urban Sustainability Transitions*, Theory and Practice of Urban Sustainability Transitions,
DOI 10.1007/978-4-431-55426-4

Theory and Practice of Urban Sustainability Transitions

This book series Theory and Practice of Urban Sustainability Transitions is intended to explore the different dynamics, challenges, and breakthroughs in accelerating sustainability transitions in urban areas across the globe. We expect to find as much different and diverse stories, visions, experiments, and creative actors as there are cities: from metropolises to country towns, from inner city districts to suburbs, from developed to developing, from monocultural to diverse, and from hierarchical to egalitarian. But we also expect to find patterns in processes and dynamics of transitions across this diversity. Transition dynamics include locked-in regimes that are challenged by changing contexts, ecological stress and societal pressure for change as well as experiments and innovations in niches driven by entrepreneurial networks, and creative communities and proactive administrators. But also included are resistance by vested interests and sunken costs, uncertainties about the future amongst urban populations, political instabilities, and the erosion of social services and systems of provision. And finally there are the forming of transformative arenas, the development of coalitions for change across different actor groups, the diffusion and adoption of new practices, and exponential growth of sustainable technologies.
For this series we seek this middle ground: between urban and transition perspectives, between conceptual and empirical, and between structural and practical. We aim to develop this series to offer scholars state-of-the-art theoretical developments applied to the context of cities. Equally important is that we offer urban planners, professionals, and practitioners interested or engaged in strategic interventions to accelerate and guide urban sustainability transition frameworks for understanding and dealing with on-going developments, methods, and instruments.
This book series will lead to new insights into how cities address the sustainability challenges they face by not returning to old patterns but by searching for new and innovative methods and instruments that are based on shared principles of a transitions approach. Based on concrete experiences, state-of-the-art research, and ongoing practices, the series provides rich insights, concrete and inspiring cases as well as practical methods, tools, theories, and recommendations. The book series, informed by transition thinking as it was developed in the last decade in Europe, aims to describe, analyse, and support the quest of cities around the globe to accelerate and stimulate such a transition to sustainability.

To sum up, the book series aims to:

- Provide theory, case studies, and contextualized tools for the governance of urban transitions worldwide
- Provide a necessary and timely reflection on current practices of how transition management is and can be applied in urban contexts worldwide
- Further the theorizing and conceptual tools relating to an understanding of urban sustainability transitions
- Provide best practices of cities across countries and different kinds of cities as well as across policy domains in shaping their city's path towards sustainability

More information about this series at http://www.springer.com/series/13408